T0181409

The Internet Challenge: Technology and Applications

The Internet Challenge: Technology and Applications

Proceedings of the 5ᵗʰ International Workshop
held at the TU Berlin, Germany, October 8ᵗʰ-9ᵗʰ, 2002

Edited by

GÜNTER HOMMEL
Technische Universität Berlin,
Berlin, Germany

and

SHENG HUANYE
Shanghai Jiao Tong University,
Shanghai, China

Springer Science+Business Media, B.V.

A C.I.P. Catalogue record for this book is available from the Library of Congress.

ISBN 978-94-010-3942-0   ISBN 978-94-010-0494-7 (eBook)
DOI 10.1007/978-94-010-0494-7

Printed on acid-free paper

Contents

vi

Contents

Workshop Co-Chairs

Günter Hommel, Technische Universität Berlin
Sheng Huanye, Shanghai Jiao Tong University

Program Committee

Martin Buss
Kurt Geihs
Sergei Gorlatch
Hans-Ulrich Heiß
Günter Hommel (Chair)
Sheng Huanye
Adam Wolisz
Fritz Wysotzki

Organizing Committee

Wolfgang Brandenburg
Michael Knoke

Workshop Secretary

Gudrun Pourshirazi
Silvia Rabe
TU Berlin
Institut für Technische Informatik
und Mikroelektronik
Einsteinufer 17
10587 Berlin
Germany

Preface

The International Workshop on "The Internet Challenge: Technology and Applications" is the fifth in a successful series of workshops that were established by Shanghai Jiao Tong University and Technische Universität Berlin. The goal of those workshops is to bring together researchers from both universities in order to exchange research results achieved in common projects of the two partner universities or to present interesting new work that might lead to new cooperation.

The series of workshops started in 1990 with the "International Workshop on Artificial Intelligence" and was continued with the "International Workshop on Advanced Software Technology" in 1994. Both workshops have been hosted by Shanghai Jiao Tong University. In 1998 the third workshop took place in Berlin. This "International Workshop on Communication Based Systems" was essentially based on results from the Graduiertenkolleg on Communication Based systems that was funded by the German Research Society (DFG) from 1991 to 2000. The fourth "International Workshop on Robotics and its Applications" was held in Shanghai in 2000 supported by VDI/VDE-GMA and GI.

The subject of this year's workshop has been chosen because both universities have recognized the fact that internet technology is one of the major driving forces for our economies now and in the future. Not only the enabling technology but also challenging applications based on internet technology are covered in this workshop. The workshop covers the scope from information extraction, data analysis, e-learning and e-trading over the areas of robotics, telepresence, communication techniques and ends with metacomputing, electronic commerce, quality of service aspects and image retrieval.

At TU Berlin the German Research Society (DFG) has been funding a new Graduiertenkolleg on "Stochastic Modeling and Analysis of Complex Systems in Engineering" since 2000. Results from this Graduiertenkolleg but also from other projects funded by different institutions are presented in this workshop. The workshop is supported by the special interest group of GI and ITG "Communication and Distributed Systems". Financial support by DAAD for the bilateral exchange of scientists between our universities is gratefully appreciated. We also gratefully recognize the continuous support of both universities that enabled the permanent exchange of ideas between researchers of our two universities.

Berlin, June 2002
Günter Hommel, Sheng Huanye

AN INTERNET BASED MULTILINGUAL INVESTMENT INFORMATION EXTRACTION SYSTEM

Fang Li, Huanye Sheng, Dongmo Zhang, Tianfang Yao
Dept. of Computer Science, Shanghai Jiao Tong University, Shanghai, 200030, China

li-fang@cs.sjtu.edu.cn

Abstract: Electronic information grows rapidly as the Internet is widely used in our daily life. People can easily obtain information from the Internet. In order to identify some key points from investment news automatically, a multilingual investment information extraction system is realized based on templates and patterns. The system consists of three parts: user query processing, extraction based on templates and patterns and dynamical acquisition. The system features the uniform processing for different languages and the combination of predefined templates and dynamic generated templates. Currently the system processes queries in Chinese, English, German and extracts Chinese investment news from the Internet, German and English investment news will be added in the future.

Key words: Information Extraction, Internet Application

1. INTRODUCTION

Internet has become an important part of our daily life. It hosts huge amount of information related to entertainment, scientific discovering, information acquisition, electronic business and so on. The ever-spreading tentacles of the Internet have revived the research on multilingual information processing. How to get information quickly and accurately

1

G. Hommel and S. Huanye (eds.), The Internet Challenge: Technology and Applications, 1–9.

without any language barriers from the Internet is the challenge for many scientists and experts on different research areas. Because of the complexity of natural languages, accurate information retrieval and robust information extraction still remain tantalizingly out of reach.

Information extraction is the process of identifying relevant information where the criteria for relevance are predefined by the user in the form of a template. Below is a passage of investment news in Chinese:

本报讯，世界最大的芯片制造巨头英特尔宣布增加在华投资。该公司宣布向位于上海浦东外高桥的生产制造企业新增投资 3.02 亿美元，使其在上海的封装/测试厂投资总额达到 5 亿美元。这次新追加的投资将引进技术和设备，用于验证、测试和封装最新的支持英特尔奔腾 4 处理机平台的英特尔 845 芯片组[1]。

The filled template corresponding to the above news is:
Company name:英特尔
Company to be invested: 生产制造企业
Its place: 上海浦东
Newly invested money: 3.02 亿美元
Amount of invested money: 5 亿美元
Currency: 美元
Content of investment: 用于验证、测试和封装最新的支持英特尔奔腾 4 处理机平台的英特尔 845 芯片组

Even if Information Extraction seems to be a relatively mature technology, it still suffers from a number of unsolved problems that limited the application only on the predefined domain. Many effects have been focused on such issues [1][2][3].

In this paper, we introduce an investment information extraction system, which is oriented to the multilingual information extraction. German, English and Chinese have applied in this experimental system. In the following, the system architecture is first described, then, some features of the system are introduced, such as predefined templates and dynamic acquired templates, language-independent templates and language-dependent patterns. Finally some evaluations and conclusions will be given.

---

[1] Translation: News report, The world biggest chip maker INTEL company has announced that it will invest another 302 million US$ to the Production and Manufactory enterprise situated in Shanghai Pu Dong economic zone. The amount of investment in Shanghai INTERL package and test factory has reached 500 million US$. The newly appended money will be used in buying new devices and technologies for evaluation and test the new 845 chips supporting for Pentium 4 process platform.

2. SYSTEM ARCHITECTURE

The system (shown in figure 1) consists of three parts: user query processing, extraction based on templates and patterns and dynamical acquisition:

- User query processing provides two possibilities for the users. One is template-based keyword. The other is natural language question, such as who invested in some company? Which company has been invested by INTEL company? There are question templates defined in the system. If the user inputs other questions, the system picks up the most similar question to process. All the queries can be made in one of the three languages: Chinese, English and German.
- Extraction module receives a user query and searches the tagged corpus for relevant contents according to predefined templates and patterns. The templates are language independent, it is defined by the event in reality. However patterns describe the extraction rules related to each slot, they are different from one natural language to another.
- Dynamic acquisition extracts the templates and the patterns from the tagged corpus to complement the predefined templates made by human being. It makes the extraction system easy to adapt to other domains.

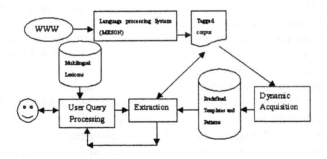

Figure-1 Architecture of the system

4

The corpus is collected from financial news on the Internet, it is tagged by a shallow parsing system called MESON[2], with English and German shallow parsing ability. For Chinese, It integrates a modern Chinese Automatic Segmentation and Part-of-Speech (POS) Tagging System[3], some Chinese grammar rules[4] for company name, person name, money, dates and other name entities (NE) are added in the MESON. The interface of the system is in the figure 2. It is realized by Java and XML.

Figure-2 Question-Interface of the system

Comparing with other information extraction systems, the experimental system has some features in the following:
1. Integrating some components already developed instead of starting from scratch. The system consists of components of different functions, either borrow from others or develop by ourselves.
2. Combining static information with dynamic information, such as predefined templates and dynamic acquired templates, template-based keywords and question query in user query processing.
3. Focusing on multilingual texts to realize a multilingual information extraction based on language independent templates and language dependent patterns. Therefore, uniform processing can be achieved for different languages. In the following, some features are detailed.

---

[2] MESON was developed by Markus Beck in German Research Center for Artificial Intelligence (DFKI), Saarbruecken, Germany.
[3] Modern Chinese Automatic Segmentation and POS Tagging System was developed by Shan Xi University in CHINA.
[4] Written by Edith Klee and Tannja Scheffler from Saarbruecken University.

3. PREDEFINED TEMPLATES AND DYNAMIC ACQUIRED TEMPLATES

In the system, we first defined two templates about the investment event by hand. One template is about the normal activity of investment such as the example in Introduction, the other is about the investment on stock market, which concerns shares, the unit price and so on. Through dynamic acquired process, we can get the templates about merge, acquisition and so on, as long as texts describing such events appear in the corpus. The process is described in the following:

1. The user inputs an event name, all examples about this event in the corpus will be identified.
2. Show the first example with identified actors of the event, time of the event, location of the event and so on, i.e. define slots of the template, and also the type of slots according to the POS or NE automatically.
3. User can make some corrections on those identified slots. The correction is made on all of the examples related to this event automatically.
4. Show the next example, and do step 2 and step 3.
5. Repeat until the final template is correct and complete.

In fact, more examples will give the new template a wide coverage, it needs of cause more human help. On the other hand, the template can be generated automatically without user interaction, however, the precision is unsatisfied. We should balance between the two factors. For example, there is an event: acquisition. The news report is as following:

Isaac 公司将以换股形式收购软体公司, 收购总价格为 8 亿 1000 万美元[5]

After tagging, the passage became:
[FIRM-NP Isaac 公司] ("将" ("将" NIL . :D)) ("以" ("以" NIL . :P)) ("换" ("换" NIL . :V)) [NP 股 形式] ("收购" ("收购" NIL . :V))[FIRM-NP HNC 软体公司] ("，" ("，" NIL . :W)) [NP 收购 总 价格] ("为" ("为" NIL . :P)) [MONEY 8 亿 1000 万美元] ("。" ("。" NIL . :W))

The system identifies the actor of the acquisition is Isaac 公司 (Isaac company), the type of the actor is FIRM-NP, the company to be acquired is HNC 软体公司 (HNC soft company) and its type is also FIRM-NP. The money of acquisition is: 8 亿 1000 万美元 (810 million). From the above example, the system generates three slots for this event, after analyzing the second example, the system may find another 2 slots, such as the place of the

---

[5] Translations: Isaac company will acquire the HNC company by stock holding. The amount of the acquisition is 810 million US$.

acquisition, and the date. The slots will be more and more complete as the system analyzing more examples. During the whole process, a user can make correction before the template is generated. Finally the system generates the new template of acquisition event with XML format in the following:

```
<template event="acquisition" id="t1001">
<slot id="s1001001" name="actor of the acquisition" type="FIRM-NP">
<slot id="s1001002" name="acquired company " type="FIRM-NP">
<slot id="s1001003" name="amount of money" type="MONEY">
<slot id="s1001004" name="acquisition date" type="DATE">
<slot id="s1001005" name="place of the acquisition" type="PLACE">
<slot id="s1001006" name="the percentage of acquisition " type="M">
<slot   id="s1001007"   name="country   of   the   acquired   company"
type="COUNTRY">
```

4. LANGUAGE-INDEPENDENT TEMPLATES AND LANGUAGE-DEPENDENT PATTERNS

In order to realize a multilingual information extraction system, templates combing patterns is our solution to solve the multilingual problems. The structure of the templates and patterns is in the following:

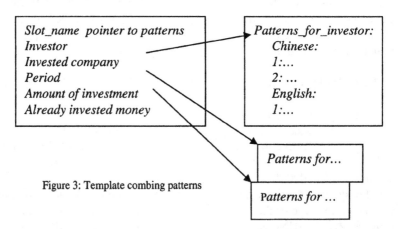

Figure 3: Template combing patterns

Templates and patterns are written in XML in order to have unified resources and platform-independent realization. For example, some patterns of Chinese and English for the actor of investment is listed in the following:

```
<template id="t001" event="investment">
    <slot id="s001001" name="investor" type="FIRM-NP">
        <pattern language="Chinese">
          <stuff type="FIRM-NP"/>
          <substitutable>投资</substitutable>
          <phrase type="MONEY"/>
        </pattern>
        <pattern language="English">
          <stuff type="FIRM-NP"/>
          <select>
             <li>are</li>
             <li>is</li>
          </select>
          <substitutable>interested in</substitutable>
          <substitutable>investing in</substitutable>
          <select>
             <li>
                <phrase type="PLACE"/>
             </li>
             <li>
                <phrase type="FIRM-NP"/>
             </li>
          </select>
        </pattern>
```

In the above, *stuff type* means the type of the slot, *substitutable* means the word can be substituted by its synonym or other morphological forms, *select* means one of them can be selected, *phrase type* means the type of a phrase followed, *removable* means the word may be removed, *notext* means there is no text between two words. With those parameters, we can describe all kinds of patterns in multilingual form.

5. SYSTEM EVALUATION AND ANALYSIS

The training corpus consists of 80 news reports collected from the Internet. There are about 400 sentences in the corpus. The test corpus consists of 50 news reports. The test result for extraction is shown in the table 1. Let *all* be the total number of extracted slot content, *Act* the number of correct and *false* number of wrong answer, *miss* the number of omitted one. We define the precision is Act/all, recall is Act/ (all+miss).

Table -1. The result of extraction

	Precision	Recall	P & R
Investor name	83%	76%	80%
Invested name	64%	58%	61%
Location of the investor	69%	69%	69%
Location of the invested	88%	88%	88%
Stock buyer	100%	90%	95%
Stock seller	100%	75%	88%
Amount of money	90%	77%	83%
date	100%	89%	94%

The performance is OK according to the state-of-the-art and to the time spent on development. However, we analyze the remaining errors as following:

• Some errors are caused due to the absence of reference resolution and discourse analysis. At the moment, in our system, there is no co-reference resolution and discourse analysis, the system cannot recognize the real entities of pronouns, such as, it, this and so on. Some key information is across the sentence boundary. Therefore, some slots has been omitted or misunderstood.

• Some errors are caused by the diversities of natural languages, the complexity of the order. The more patterns, the higher the precision.

For the template dynamic generation, according to the test, we find that without human corrections, the precision is 85.27%, the recall is 78.01%, if a user makes corrections on 5 examples, the precision is 88.55%, the recall 82.27%, if the user makes corrections on 10 examples, the precision and recall remain the same, if the user corrects more examples, for example, 15 or 20 examples, the precision is 89.31%, and the recall is 83.57%. Therefore, a user makes corrections on 5 examples, both the precision and the recall will rise 3% or so. Not too much human involvement is needed in the dynamic generation.

CONCLUSION

In this paper, we describe the multilingual investment information extraction system based on templates and patterns. Currently the user queries can be in Chinese or English or German. But the tagged corpus is only in

Chinese. However, English and German corpus can be appended without revision of the system except the addition of patterns for English and German. Using templates and patterns to realize a multilingual information extraction has its advantage and disadvantage. The advantage is unified realization for different languages. The disadvantage is that patterns cannot have a wide coverage of news reports. On the other hand, extraction relies on the analysis of natural language processing, i.e. the POS or NE. How to extract information from the raw text is our challenge in the future.

ACKNOWLEDGEMENTS

We are grateful to the people in DFKI, especially to Mr. Markus Beck, Miss. Edith Klee and Miss. Tannja Scheffler for their cooperation and helpful discussion. We also thank Mr. Miao Wang, Mr.Cheng Zhou and Mr. YiDa Zhang for the programming of the experimental system. At last, we have to mention that the China Natural Science Foundation has supported the research work under the grant no: 60083003.

REFERENCES

1. Thierry Poibeau, " Deriving a multi-domain information extraction system from a rough ontology" IJCAI-2001 pp.1264-1269
2. Chia-Hui Chang, ShaoChen Lui, " IEPAD: Information Extraction Based on Pattern Discovery" WWW10 on May 1-5, 2001 HongKong
3. Christopher S.G.Khoo, Syin chan, Yun niu, "Extracting Causal knowledge from a Medical Database Using Graphical Patterns" in Proceedings of the 38th Annual Meeting of the Association for computational Linguistics 1-8 Oct. 2000 HongKomg. pp.336-343

AN INTELLIGENT SEARCH ENGINE
USING IMPROVED WEB HYPERLINK
MINING ALGORITHM

Zhang Ling, Ma Fanyuan and Ye Yunming
Dept of Computer Science and Engineering, Shanghai JiaoTong University, Shanghai 200030*
fyma@mail.sjtu.edu.cn

Abstract   This paper presents a new search engine-*Luka*. *Luka* uses the cluster crawlers to enhance the search performance. We put forward a new web page ranking algorithm to improve the relevance of the results of search engines in *Luka*. This algorithm combines a link analysis algorithm with content correlation techniques to dynamically rank the pages, which are more relevant to the query. We use the anchor text of a Web page to compute its similarity to the query and use the similarity to readjust the rank of the page pre-calculated with the PageRank algorithm. Our experiments have shown a significant improvement to the current link analysis algorithm.

Keywords: crawler, content correlation analysis, hyperlink analysis,search engine

1.   Introduction

People use search engines to find information on the World Wide Web. In general, a search engine usually consists of web crawler, document indexer and user interface. The high performance, robust web crawler and efficient index (or ranking) algorithm are the key factors to the quality of a search engine. In this paper, we introduce a distributed web crawler system, which is implemented using Java programming language. The optimized data structure design and the communication coordination between machines fulfill high efficient web page downloading and URL management. And we also propose a smart index algorithm that considers both the content similarity evaluation and hyperlink mining algorithm and compare it with PageRank [1]. The experiment shows that this algorithm improves the web query precision for site search. The rest of this paper is organized as follow: Section 2 discusses the design details of a distributed web crawler and its Java implementation. Section 3 reviews the

*Funding provided by NSFC Major International Cooperation Program NO.60221120145.

G. Hommel and S. Huanye (eds.), The Internet Challenge: Technology and Applications, 11–18.
© 2002 *Kluwer Academic Publishers.*

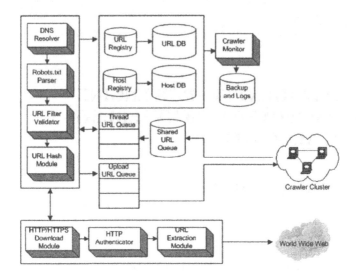

Figure 1.   Web crawler system architecture

current web information retrieval algorithms and discusses the drawbacks of each method. Section 4 introduces the improved ranking algorithm CALA. The last two sections are experiments and conclusion.

2.   Distributed Web Crawlers

Web Crawler traverses the WWW and downloads web pages recursively. We've implemented a Java based distributed web crawler to download web pages and index them. The data structures for URL storing in the crawler are optimized for CALA.

The system architecture of the Web crawler can be seen in Figure 1.

When the Web crawler starts, it resolves the URL to IP address and checks whether this URL has already been downloaded. If it is a new URL, the crawler checks whether it is access allowed by parsing the robots.txt file on the Web server, then this URL is checked subsequently by the URL filters. If the URL belongs to the local machine download job after hash computing, the crawler will download it and extract the hyperlinks from the web page, then add them to the URL queue; else the URL will be inserted to the upload URL queue.

The crawler cluster consists of several machines with one coordinator. The coordinator machine can communicate with the crawler machine using Java RMI. The peer to peer communication between two individual machines is forbidden, and they swap URLs through the coordinator. The primary job of the coordinator is collecting and dispatching URL groups to different machines according to their routing destinations.

The Web crawler deals with millions of URLs and will encounter duplicated URLs when crawling the Web. There are nearly 8.5% duplicated URLs when traversing the WWW [2], and it will bring extra cost if we don't judge whether a URL has already been crawled.

For detecting duplicated URL, different crawler systems use different schemes. The WebCrawler from Washington University applies commercial databases storing URLs [3]; Mercator, which comes from Compaq Research Center, generates a 64 bits checksum for each document using the fingerprinting algorithm. Mercator performs an interpolated binary search of an in-memory index of the disk file to identify the disk block on which the fingerprint would reside if it were present. It then searches the appropriate disk block, again using interpolated binary search.

To implement fast URL searching, we use memory based URL storage and search approach. All the URLs that have been visited are stored in a Trie [4] structure, which resides in memory. Searching Trie object is faster than querying RDBMS and other memory/disk swapping indexing methods. For Java implementation of Trie data structure, the average URL searching time is about 0.24 ms.

It is not practical for single stand-alone crawler using memory for URL indexing, because the crawler will soon be out of memory. Since our distributed Web crawlers can be load-balanced by adding more machines, the average hardware resource cost of a single machine can be kept at a lower level.

3.   Web Information Retrieval algorithms

3.1   Vector Space Model

Content correlation analysis computes the similarity between the query and destination documents. This method is widely used in the traditional search engine to determine whether a document is relevant to a given query. Before discussing CALA, we will first introduce Vector Space Mode (VSM) [5], with which we compute the content correlation between two documents.

VSM, a mathematical mode to compute document-matching degree, is applied in information retrieval. In VSM, a vector denotes the document or the query. The matching degree of the two documents is measured by the cosine of the angle between their two vectors.

$$Similarity(D, Q) = \frac{\sum_{i=1}^{n} d_i q_i}{\sqrt{\sum_{i=1}^{n} d_i^2 \sum_{i=1}^{n} q_i^2}} \tag{1}$$

Though VSM can match the user query and target documents in some degree, but it only considers the document content and web page designers can insert hidden text to promote the ranking in the search results.

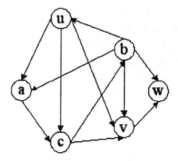

Figure 2.   directed link graph G

3.2   Web Hyperlink analysis

The basic idea of the link analysis algorithm is as follows:

We can model the Web (or page set) as a graph. The graph contains a node for each page in the page set and there exists a directed edge (u,v) if and only if page u contains a hyperlink to page v. We call this directed graph the link graph G.

In Figure 2, node b, c and u contributes to the rank weight of node v because they all contain the hyperlink to node v. The more hyperlinks points to a page, the more important the page is. The drawback of this approach is that each link is equally important and neglects the quality of the source page. In most situations, high quality web pages often contain hyperlink to high quality pages.

To solve this problem, S.Brin and L.Page proposed PageRank algorithm. The PageRank of a web page is computed by weighting each hyperlink proportionally to the quality of the page containing the hyperlink. The PageRank of a page can be defined as follow:

$$PR(v) = \varepsilon/n + (1 - \varepsilon) \sum_{(u,v) \in G} PR(u)/outlink(u) \qquad (2)$$

where

ε is a dampening factor usually set between 0.1-0.2;

n is the number of nodes in G;

outlink(u) is the number of hyperlinks in page u.

We can compute the PageRank of each page in Figure 2, and the results are in Table 1(ε=0.2):

PageRank algorithm converges fast. Iterated computation on Figure 2 converged after 15 times of iteration. In practice roughly 100 iterations suffice [7].

PageRank provides an efficient way to rank Web resources, and improves the precision of information retrieval [8]. However, PagePank is a query-

Table 1.   PageRank value of Figure 2

	Node a	Node b	Node c	Node u	Node v	Node w
PR	0.060210	0.071004	0.094177	0.047534	0.097881	0.125839

independent ranking algorithm and the weight of each web page is computed only once. In fact, a web page that has a high PageRank will not always be valuable according to different search keywords. For example, a Web page contains much valuable information on data mining has been linked to many famous data mining research institutes, so it has a high PageRank. Unfortunately, this page also includes some piece of information on data compression, so when a query "*data compression*" is submitted to the search engine, this document may appear on the top of the search results only because the page both includes this keyword and has a high PageRank. The search engine that only adopts static link analysis algorithm may produce misleading search results.

4.   CALA

The main reason to use link analysis instead of content similarity computing is to avoid artificial disturbance in the Web pages. To keep the objective criterion, we do not use the Web page content for content correlation computing.

In fact, there is another important candidate to be used for computing content similarity: the anchor text. Hyperlink in the Web pages uses anchor text describing what topic the hyperlink is about. For example:

"<a href="http://ecom.sjtu.edu.cn:8888/luka.html">LUKA Search</a>" is an HTML code fragment that contains hyperlink and its anchor text. Here the anchor text is *LUKA Search*. Anchor text reflects the comment of the author who adds the link in the web page. The comment or recommendation is independent of the linked-in page content and its author. Our new algorithm dynamically computes the similarity between user query and anchor text that links to the matched documents, and we consider the computing result as weighted factor in the link analysis function. Then we will have an improved algorithm taking into account both the link and content information.

With VSM to compute content correlation, we have a ranking criterion based on the anchor texts. CALA implements an integrated algorithm considering both the link analysis and the content similarity.

CALA can be defined as follow:

$$CALA(d) = PR(d)\{1 + \delta \cdot Similarity(d, Q)\} \qquad (3)$$

where PR (d) and Similarity (d, Q) are (1) and (2) respectively.

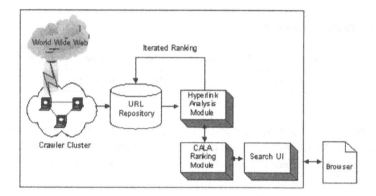

Figure 3. *Luka* search engine components

δ is the weighted factor. δ indicates the importance of content similarity in the rank computing. According to our experiments, δ is usually set to less than 0.4.

(3) shows that even if the anchor text doesn't contain valuable information about the linked document, for example, the URL string is used as the anchor text, CALA algorithm can still work with the basic link analysis algorithm.

In Section 5, we will introduce a search engine prototype that build on the CALA and test the search quality of that prototype.

5.   Experiments

5.1   Luka Intelligent Search Engine

We   have   designed   a   search   engine-*Luka* (http://ecom.sjtu.edu.cn:8888/luka.html).   The main components in *Luka* can be seen in Figure 3.

Luka takes advantage of CALA ranking algorithm, and performs dynamic ranking based on user query. The CALA value of each URL is displayed in the search result page by descending order.

URL Repository is a database resides in disk, and it contains URL, HTML title, hyperlink structures and anchor text, etc. The hyperlink analysis module runs offline, and writes the PageRank value of each URL to the URL Repository after iterated ranking.

5.2   Web Crawler Testing

To evaluate the performance of the distributed Web crawlers, we build a small cluster, which consists of five machines (Pentium III 550MHz with 256MB memory). These machines are installed with Microsoft Windows and Sun JDK

1.3. Four of them run as Web crawler, and the other one runs both as a Web crawler and a coordinator. The internal network of the cluster is 100Mbps, and they are connected to the Internet through a 10Mbps gateway.

We choose the homepages of twenty famous universities in China as the initial URLs for breadth-first crawling. Before the test begins, we restrict the maximum single file length to 500KB and total download size for each crawler to 2GB. Table 2 is the statistic data after the crawler cluster running for 7 hours.

Table 2.   Web Crawler Performance Statistic

URL Registry	1,317,178 [URL]
URL Download Success	863,476 [URL]
URL Download Fail	265,389 [URL]
Average download speed	34.43 [URL/s]; 333.81 [KB/s]

[6] mentions that each crawler of Google search engine keeps roughly 300 connections open at once. In the average, the system can crawl over 33.5 web pages per second using four crawlers. In our experiments, each crawler creates 5 threads for concurrently crawling. So there are total 25 connections open at once, while Google has 1200 connections.

5.3   Web querying evaluation

Currently, *Luka* search engine only covers the web pages in CERNET (China Education and Research Network), so the query terms we choose are all related with the education in China. These keywords include English and Chinese words, such as university names, journal titles and some abbreviations.

We consider that *Luka* is especially suitable for site search because we do not index the full text but only the HTML title and hyperlink anchor text of the Web page. Since CALA improves search quality by adopting both structure and content information, it can find highly related web sites even the keywords don't appear in the web pages. For example, we search our URL repository using "*sjtu*", "*fudan*", "*tsinghua*" and "*pku*", which are English names or abbreviations of four famous universities in China. The top web sites in the search results are the homepages of these universities. These homepages are all Chinese version, and the query keywords don't appear in these homepages. If we use the Chinese abbreviations of these universities, the top web sites in the search results remain the homepages of these universities.

6. Conclusions

In this paper, we introduce the design and Java implementation details about a distributed Web crawler. The improvements in data structure and URL queue designs make the crawling fast and effective. Designing a Web crawler is really a non-trivial task both designers and developers. From our current research experiences on Web crawler, data structure, memory management, bottle neck analysis, crawling policies, current specifications on WWW and many other problems need to be studied.

To improve Web search quality, we proposes a new ranking algorithm CALA that combines HTML anchor text and hyperlink structure, and we develop a search engine based on our algorithm. The experiments show that CALA improves the search quality of Web search. This paper also describes the design and implementation details of a distributed Web crawler system, which achieves high speed Web crawling and high efficient URL management.

Luka intelligent information retrieval system provides a basic platform for Web mining research, and the distributed Web crawler enables other research works such as Web information extraction, semi-structured data management and other related works.

References

[1] Taher H. Haveliwala. *Efficient computing of PageRank.* Stanford Database Group Technical Report, 1999.

[2] A. Heydon and M. Najork. Mercator: *A scalable, extensible Web crawler.* World Wide Web, 2(4): 1999, page: 219-229.

[3] Brian Pinkerton. *WebCrawler: Finding What People Want.* Ph.D dissertation, University of Washington, 2000.

[4] E. Fredkin. *Trie memory.* Comm. ACM, Vol. 3, 1960. 26, page: 490-500.

[5] S.K.M.Wong, W.Ziarko, V.V.Raghavan. *On modeling of information retrieval concepts in vector spaces.* ACM Transactions on Database Systems, Volume 12, Issue 2, 1987, page:299-321

[6] S.Brin and L.page. *The anatomy of a large-scale hypertexual web search engine.* In Proc.of the WWW7 Conference, Brisbane, Australia, April 1998, page 107-117.

[7] Monika Henzinger. *Link Analysis in Web Information Retrieval.* IEEE Data Engineering Bulletin, September 2000, page 3-8.

[8] S.Lawrence and C.L.Giles. *Accessibility of information on the Web.* Nature, 400, 1999, page 107-109.

DATA ANALYSIS CENTER BASED ON E-LEARNING PLATFORM

Ruimin Shen, Fan Yang, Peng Han
Department of Computer Science, Shanghai Jiaotong University, Shanghai, China

Abstract: Web based learning enables many more students to have access to a distance-learning environment, providing students and teachers with unprecedented flexibility and convenience. At the same time, current E-Learning systems also pose many problems. For example, teachers can't know the learning status of students, the course content is static and the teacher's assignment given to every student is the same. In this paper, we present a Data Analysis Center based on E-Learning Platform that solves these problems. In particular, our system can show the place of a student in the class, the weakness during learning process. Furthermore, it can help the teacher to analysis students' learning patterns and organize the web-based contents efficiently. The system is intelligent through data mining features and user-friendly through visualized services to both teachers and students.

Key words: Data Analysis Center, Web Mining, e-learning

1. INTRODUCTION

As distance learning becomes one of the hotspots in network research and application, many web-based education systems have been established all around the world, such as the Virtual-U[1] and Web-CT [2]. To cover all the phases of the learning process, these systems are usually comprised of such fundamental components as synchronous and asynchronous teaching system, course-content delivery tools, polling and quiz modules, virtual workspaces for sharing resources, white boards, grade reporting systems, assignment submission components, etc. These research and products enable large groups of dispersed individuals to interact, collaborate and study over the Internet.

G. Hommel and S. Huanye (eds.), The Internet Challenge: Technology and Applications, 19–28.
© 2002 *Kluwer Academic Publishers.*

But current distance learning systems still have some shortcomings: firstly, the students learn through the Internet, so their learning status can not be collected by systems and teachers can't know if the progress of the course is fitful, how his students learnt, what they want to learn further, and so on. All of these will lead to the gap between students and teachers. Secondly, the course content is static and the teacher's assignment and content given to every student is the same. In reality, students have different background and the knowledge structure is dynamic. Given such diversity, how do we analyze students learning habits, characteristics and knowledge structure? In addition, how to visualize the analysis results to teachers and students more intelligible?

In order to solve these problems, our system proposes an overall subsystem ---the Data Analysis Center, which includes an analysis tool to support the system.

In the next section we present a brief introduction to the Data Analysis Center focusing on the architecture and key functions. Section 2 presents the architecture of the Data Analysis Center.

The function and key technologies will be discussed in Section 3. In Section 4, we will give a case study of it. We conclude our discussion in Section 5, where we will also explore our future work.

2. THE ARCHITECTURE OF THE DATA ANALYSIS CENTRE

It is a very difficult and time consuming task for the educators or the e-learning web site sponsors to thoroughly track all the activities performed by learners. So the educators are in desperate need for non-intrusive and automatic ways to get objective feedback from learners in order to better follow the learning process and appraise the on-line course. On the learner's side, it would also be very useful if the system could automatically guide the learner's activities and intelligently recommend on-line activities or resources that would favor and improve the learning.

Web mining is the use of data mining techniques to automatically discover and extract information from Web documents and services [9, 10, 11, 12]. It is now the main technology used in e-commerce to make the web sites more adaptive and personalized and hence more attractive to customers. Up to now, many commercial systems have been developed using this technology, for example, systems for recommendation on Amazon.com that suggests books to purchase related to a current purchase based on preference information and similar user purchases. The same concept and technology can also be used on e-learning.

Unlike in commercial systems, which mainly use web log files as their data analysis source, there are much more information in e-learning environment that should be used to analyze, such as the learning progress, the assignment completion, the admitted question, the exam score and so on. We want to build a Data Analysis Center based on data mining and web mining technologies to deal with learning information so as to construct a dynamic/intelligent/personalized distance learning environment. The whole system comprises of three main parts: didactical information warehouse, data mining engine and visualized analysis tool. The framework is shown as Figure 1.

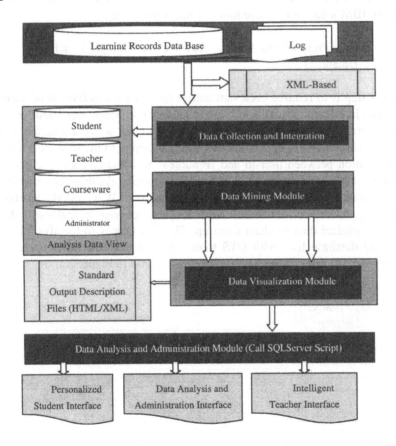

Figure 1. The Framework of the Data Analysis Center

3. THE FUNCTION AND KEY TECHNOLOGIES OF DATA ANALYSIS CENTRE

3.1 Construct Didactical Information Warehouse

In order to process these learning information and extract valuable patterns that could be used to enhance the learning system or help in the learning evaluation, a signification pretreatment and warehouse-building phase needs to take place so as to prepare the information for data mining algorithm. In this phase we will establish a didactical information warehouse with the IBM's data warehouse tool (Visual Warehouse).

3.1.1 Data Pretreatment And Integration From Multiple Function Databases

First step is to remove irrelevant entries based on the definition of useful attributes in e-learning platform. Since a student can access our learning site everywhere, so a user-id maybe correspond many IP address which would lead to error in session data collection. Second step is to identify the one-to-multi relation between user-id and IP address with their access log entries and Id-IP tables in our source database. The resource database is composed of two kinds of data: the log file with specification of W3C and the attribute tables in sub-function database. The data-preprocessing module will deal with the original data to clean them up. The first task is to transfer the log files into database files with DTS tools. The second task is to create the corresponding table of the User_ID and IP as shown in Figure 2.

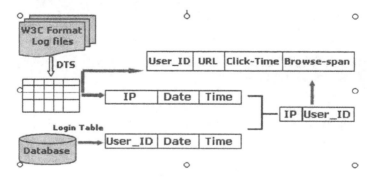

Figure 2. Log Data Preprocessing Process

Such operation can deal with the problem of one-to-many relation between student's User_ID and IP attributes. The third task is to calculate the click-time and browse-span of one URL, which is very important to mining the data structure of students. The last task is to create some new tables and views for further analysis process.

In the third step, we plan to identify access sessions. The aim is to recognize sequences of events such as A⇒B⇒C⇒ B⇒D... where A, B, C, D, etc. are page or script requests. The challenge is to recognize the beginning and the end of sessions. In the following step, we want to map access log entries with actual learning activities. The result is a sequence of learners' relevant on-line activities of the form: Login⇒ExerciseList⇒SubmissionQuiz1⇒ExerciseList⇒ ReadConferenceMessage... Completing the traversal paths consists of inferring cache hits and proxy meddling based on the structure of the learning site and how pages and activities are effectively linked together. Finally, to integrate the cleaned click streams with existing data about learners. These valuable and beneficial data could be the profiles of the learners, their quantitative and qualitative evaluations, etc.

3.1.2   Construct Data Warehouse In A Standard Way

Before we can process the information collected from the didactical system, we must convert it into structural format (we choose XML) which can be processed by data mining algorithm. In our system, we plan to implement this by putting an agent in each functional system. While a user access the systems, the agent automatically records the useful information and organizes it by predefined format, then write it into corresponding table in the database. All these tables constitute analysis data source. In this system, we use XML as our data specification and DB2 as database platform. The challenge technologies are: define the XML schema over all learning information, build multi-agent to collection and integrate the data, develop standard interface between the agent and the DB2, use Visual Warehouse tools to establish the warehouse.

This phase --- construct didactical information warehouse may consume 50% to 60% of the effort and resources needed for data analysis.

3.2   Data Mining Engine

In this phase, we will adapt the web mining technologies for knowledge discovery in e-learning platform. And we applied the improved web mining algorithm, IBM's data mining tool (Intelligent Miner) and IBM development tools (WebSphere) to establish an intelligent and personalized data mining engine.

After our processing procedure, we can get some useful and clean tables. Since we organize our source with knowledge point and have relation-tables of source and knowledge point, so we can assessment the knowledge point in two aspects. One is general information: to calculate the Interest Measure and the Mastery Measure of each chapter-point and knowledge-point based on the statistic data. The other is personalized information: to give the Interest Measure and the Mastery Measure of one student.

After our processing procedure, we can get some useful and clean tables. Since we organize our source with knowledge point and have relation-tables of source and knowledge point, so we can assessment the knowledge point in two aspects. One is general information: to calculate the Interest Measure and the Mastery Measure of each chapter-point and knowledge-point based on the statistic data. The other is personalized information: to give the Interest Measure and the Mastery Measure of one student.

Based on the analysis results above, we used three algorithms to find some knowledge and rules. One is using the classification algorithm to classify the students into different classes based on their learning action. According to the classification of students, the teacher can organize different course content and assign difficult level homework to each class. The second is to find the association rule of knowledge-point, and give the support and confidence of each rule. The third is to organize and map the knowledge point based on the concept map algorithm

3.2.1 Evaluate Student Learning State Based On Web-Usage Mining Technology

In order to evaluate student learning state, we must collect and process the learning information in all-around way. Based on out didactical information warehouse, we can calculate hit frequency/average/median, length and duration of sessions and other limited low-level statistical measures. There have been some data mining approaches adapted specifically for web usage mining. In our system, we would use web usage mining technologies for personalization, system improvement such as web caching, network traffic improvements, knowledge structure modification and learning intelligence. There are three technologies we plan to use: first is association rules generation to discover relationships between knowledge points in a student's learning process [3-6]. In distance learning environment, the objects could be students, sessions, pages, etc. Second is sequential pattern analysis to account the sequences of learning events and mining the learning action. In other words, the fact that a knowledge point A is requested before another knowledge point B is captured in the patterns discovered. Third is interest-measure and master-measure calculation of each knowledge points. In this phase, we would build a mathematic module to calculate the measure as a liner weight combination of question frequency, assignment inaccuracy, knowledge duration and other characteristics in learning procedure. We will adapt all these techniques to give a comprehensive assessment to a student.

3.2.2 Given The Recommendation Of Learning Based On Collaborative Filtering

This phase is the fundamental of personalized learning. The main purpose is to predict what fields a person is interested in based on other course he or she has learned, predict what web pages a person will go to next based on his or her history on the site and give the relative recommendation of learning methods, didactical materials and so on.

It consists of two steps: one is to build a weight matrix of the user and the knowledge point interest-measure based on training set and assign learning plan to every class [7, 8]. The other is to classify a given student and give the corresponding recommendation of learning. We have found the following three criteria for collaborative filtering to be important: first is the accuracy of the recommendations, second is prediction time which takes to create a recommendation list given what is known about a student, third is the computational resources needed to build the prediction models. We will measure each of these criteria in our empirical comparison and attempt to measure a student's expected utility for a list of recommendations. Since different students will have different utility functions, we will try to provide a good approximation measure across many students.

3.3   Visualized Analysis Tool

Based on above analysis results, we will give a visualized analysis tool for the user to view the result. In the teacher's end, we plan to provide statistical graph on these information: assignment complement, admitted question, exam score and so on; besides we will provide the interest-measure/mastery-measure of knowledge point or chapter. The information can help teachers to reorganize the course and adapt to better fit the needs of group or an individual. In the student's end, we plan to show the knowledge structure map of every student and sign the learning process. In each point in the knowledge map, it will show the corresponding course content and quick link of reference materials. In this part the student can receive hints from the system on what subsequence activity to perform based on similar behavior by other "successful". In the administrator's end, the high-level knowledge map and other statistical graph can be shown, which can direct the administrator to adapt the course content logical structure.

4.   A DATA ANALYSIS CENTRE CASE STUDY

We have already establish a distance learning web site (http://www.nec.sjtu.edu.cn) and carried out the assistant services of both the teachers and students with the Data Analysis Center, such as student management system, answer machine, assignment system, examination system, discussion online, etc.

When a student connects to our NEC (Network Education College) home page, he can select which chapter or section to study. During student's learning process, he can ask question, do the homework, make an online examination and importantly connect to the Data Analysis Center to see his learning process, the place in his class, the knowledge point with low mastery-measure, and so on. Figure 3 give the questions of each chapter or knowledge point, and click them it can show the further association of questions and knowledge points. Based on these analysis results, students can easily learn the frequently and important knowledge according to his classmates. Besides, teachers can know which chapter or knowledge points are the difficulty during students learning process and can answer them integrated.

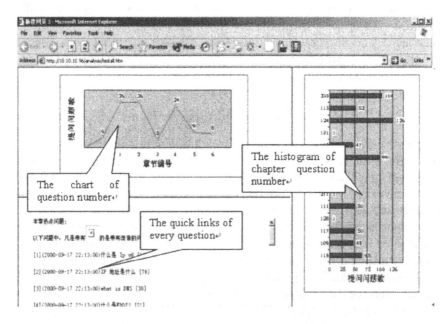

Figure 3. The Analysis Results of Questions Asked by Students

In the teacher side, just like a monitor, the analysis center also can show the online study status of students. As shown in Figure 4, the teacher can see the number of online learners, the maximum learning times, the minimum learning times and the average learning times. According to this knowledge, the teacher can know parts of the learning status of his students and can mail to the students which learning speed is too slow.

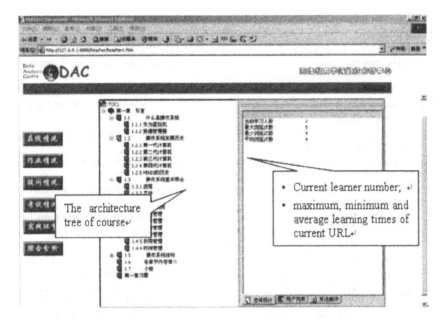

Figure 4. The Online Learning Status Monitor

5. CONCLUSION AND FUTURE WORK

In this system, we first integrate all the information of learning process by setting agent in every functional system just like a experienced teacher in a real class. Based on these multi-forms data, we propose a method to prepare and establish a data warehouse based on standard XML schema and IBM warehouse building tools. Then we propose a data mining engine to manage students, just like the CRM, based on web usage mining and collaborative filtering technologies. This system can track student's learning trace, monitor student's learning status, map student's knowledge structure, predict student's preference and so on. Furthermore, we construct multiple and intelligible visualization tools, which can help teachers to redesign their teaching plans and provide learning target recommendation and individualized course content to students.

28

In the future, we plan to offer more tests on the systems performance using the data we accumulate through real teaching sessions. Such validation will allow us to select the best intelligent teaching methods for an open, virtual teaching environment.

References
1. C. Groeneboer, D. Stockley, T. Calvert, Virtual-U: A collaborative model for online learning environments, Proceedings Second International Conference on Computer Support for Collaborative Learning; December, 1997.
   WebCT: http://www.webct.com/
2. R. Agrawal and R. Srikant. Fast algorithms for mining association rules. In VLDB'94, Sept. 1994: 487-499
3. J. Han and Y. Fu. Discovery of multiple-level association rules from large databases. In VLDB'95; Sept. 1995: 420-431.
4. R. Srikant and R. Agrawal. Mining generalized association rules. In VLDB'95; Sept. 1995: 407-419.
5. R. Srikant and R. Agrawal. Mining quantitative association rules in large relational tables. In SIGMOD'96; June 1996: 1-12.
6. J. Breeze, D. Heckerman and C. Kadie. Empirical analysis of predictive algorithms for collaborative filtering In Proceedings of the Fourteenth Conference on Uncertainty in AI, Madison, WI. 1998.
7. Sonny H.S. Chee, Jiawei Han, Ke Wang. RecTree: an efficient collaborative filtering method , Data Warehousing and Knowledge Discovery (DaWaK); Sept. 2001
8. Qiang Yang, Henry Hanning Zhang and Ian Tianyi Li. Mining Web Logs for Prediction Models in WWW Caching and Prefetching . In The Seventh ACM SIGKDD International Conference on Knowledge Discovery and Data Mining, Industry Applications Track; August 26 - 29, 2001.
9. Pitkow J. and Pirolli P. (1999). Mining Longest Repeating Subsequences to Predict WWW Surfing. Proceedings of the 1999 USENIX Annual Technical Conference.
10. Ian Tianyi Li, Qiang Yang and Ke Wang. Classification Pruning for Web-request Prediction. In Poster Proceedings of the 10th World Wide Web Conference; May 2-4, 2001.
11. Zhong Su, Qiang Yang, Hong-Jiang Zhang, Xiaowei Xu and Yu-Hen Hu. Correlation-based Document Clustering using Web Logs. In Proceedings of the 34th HAWAII INTERNATIONAL CONFERENCE ON SYSTEM SCIENCES; January 3-6, 2001.

CORRECTING WORD SEGMENTATION AND PART-OF-SPEECH TAGGING ERRORS FOR CHINESE NAMED ENTITY RECOGNITION

Tianfang Yao Wei Ding Gregor Erbach

Computational Linguistics Department, Saarland University

D-66041Saarbrücken, Germany

yao@coli.uni-sb.de wding@dfki.de gor@acm.org

Abstract: In the exploration of Chinese named entity recognition for a specific domain, the authors found that the errors caused during word segmentation and part-of-speech (POS) tagging have obstructed the improvement of the recognition performance. In order to further enhance recognition recall and precision, the authors propose an error correction approach for Chinese named entity recognition. In the error correction component, transformation-based machine learning is adopted because it is suitable to fix Chinese word segmentation and POS tagging errors and produce effective correcting rules automatically. The Chinese named entity recognition component utilizes Finite-State Cascades which are automatically constructed by POS rules with semantic constraints. A prototype system, CNERS (Chinese Named Entity Recognition System), has been implemented. The experimental result shows that the recognition performance of most named entities have significantly been improved. On the other hand, the system is also fast and reliable.

Key words: information extraction, named entity recognition, machine learning, finite-state cascades

G. Hommel and S. Huanye (eds.), The Internet Challenge: Technology and Applications, 29–36.
© *2002 Kluwer Academic Publishers.*

1.   INTRODUCTION

Information Extraction (IE) is a key language technology that aims to extract facts from documents. Since the early 90's IE technology has taken a rapid development, driven by the series of Message Understanding Conferences (MUC's) in the government-sponsored TIPSTER program [6]. It is now coming on to the market and is of great significance for information end-user industries of all kinds, especially finance companies, banks, publishers and governments [10]. As we know, named entities (NEs) are an important constituent in natural language sentences. Therefore, NE recognition (NER) is also a fundamental task of IE. In general, Chinese named entities include person name, person title, location name, organization name, product name, time, date, monetary, percentage and so on [3].

Chinese is not a segmented language, so that the words in a sentence must be segmented before they are processed by IE component. Although most papers related with Chinese IE did not deal with the relationship between word segmentation or part-of-speech (POS) tagging and the performance of IE [1], we notice that these errors have obstructed the improvement of NER performance. In order to change this situation, we propose an error correction approach for Chinese NER in this paper. Transformation-based machine learning [2, 5] is adopted in our model because it is suitable to fix Chinese word segmentation and POS tagging errors and produce effective correcting rules automatically. After using this approach, the recognition recall and precision of most named entities have apparently been enhanced.

Figure 1 is the model of our NER. The dotted line shows the flow process for the training texts; while the solid line is one for the testing texts. When training, the texts are segmented and tagged, then the error correction rules are produced and some of them are selected as the regular rules under the appropriate conditions. Thereafter, the errors caused during word segmentation and POS tagging in testing texts can automatically be corrected through utilizing such error correction rules. Among the six NEs, personal name (PN), time word (TW) and location name (LN) are recognized immediately after the error correction; while team name (TN), competition title (CT) and personal identity (PI) will be recognized by NER component.

This paper is organized as follows. Section 2 illustrates the error correction approach for word segmentation and POS tagging. Section 3 briefly gives the outline for NER component. Section 4 introduces the

---

[1] In [9] the authors have investigated the relationship between word segmentation and information retrieval.

prototype system and shows the experimental results and the appropriate analysis. Finally, section 5 draws the conclusions for this approach.

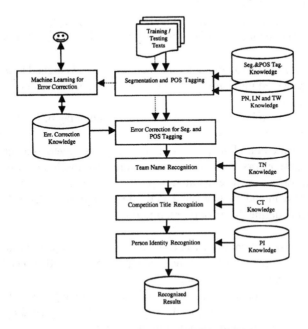

Figure -1. Named Entity Recognition Model

2. CORRECTION FOR WORD SEGMENTATION AND POS TAGGING ERRORS

For the purpose of ensuring good quality in segmenting word and tagging POS, we compared different existing Chinese word segmentation and POS tagging systems and introduced the Modern Chinese Automatic Word Segmentation and POS Tagging System [8] as the first processing component in our model. In this system, the word segmentation unit is based on the Association-Backtracking algorithm and mainly depends on Chinese language knowledge, such as word-building, form-building and syntax; while the POS tagging unit utilizes the probability statistic model as well as CLAWS, VOLSUNGA and the corresponding transmutation algorithms.

Unfortunately, we found there still exist numerous word segmentation and POS tagging errors when we make use of this system to process our texts on the sports domain. Obviously, these errors will have an effect on the consequent recognition for the NEs.

Some word segmentation and POS tagging errors related with six NE types are shown as follows:
- PN is segmented into PN and general noun or PN is tagged as a LN;
- TW and conjunction are segmented together or TW is tagged as a general noun;
- LN is segmented into general noun and verb or LN is tagged as a PN;
- TN is segmented into some parts with verb and general noun tag or TN is tagged as a PN;
- CT is segmented into some parts with general noun tag or CT keyword is tagged as a verb;
- PI is segmented into PN and general noun or PI is tagged as a verb;

Against these errors, we define some features for machine learning, such as error and correct word segmentation position, error and correct POS tag as well as context including word and POS tag for the rules. On the basis of that, we further define the error correction rule:

rectify_segmentation_error (**concat**, old_word1 | old_tag1 | old_word2 | old_tag2 | ..., concat_number, new_tag, preceding_word | preceding_tag, following_word | following_tag)

rectify_segmentation_error (**split**, old_word | old_tag, split_position1 | split_position2 | ..., new_tag1 | new_tag2 | ..., preceding_word | preceding_tag, following_word | following_tag)

rectify_segmentation_error (**slide**, old_word1 | old_tag1 | old_word2 | old_tag2 | ..., slide_direction_length1 | slide_direction_length2 | ..., new_tag1 | new_tag2 | ..., preceding_word | preceding_tag, following_word | following_tag)

rectify_tag_error (old_word | old_tag, new_tag, preceding_word | preceding_tag, following_word | following_tag)

In the error correction rule of word segmentation, **concat** means some words (or characters) that have been separated will be put together; **split** represents some words (or characters) that have been put together will be separated and **slide** denotes some words (or characters) whose word segmentation positions are not correct will be segmented newly. That is, the new position will be moved to the left or right side of the original position.

The machine learning's procedure includes detecting error positions, producing error correction rules, selecting higher-score rules, ordering rules etc. The concrete algorithm is explained as follows:
a) Compare automatic word segmentation and POS tagging with manual word segmentation and POS tagging in a sentence.

b) If they are different, record word segmentation and POS tagging environments. Otherwise transfer to f).

c) Build a new transformation rule that consists of transformation condition and action. The condition presents all triggering environment including error word segmentation position, error POS and context. The action executes correcting action that transforms word segmentation positions and POS tags.

d) Examine whether this new transformation rule is at odds with the transformation rules in the candidate rule library. If it is true, either merge rules or delete this new rule depending on both conditions. Otherwise add the new rule into the library.

e) Test the rules in the candidate rule library and choose some higher-score transformation rules that can reduce more errors. Then determine whether they are added into the final rule library.

f) If there is still sentence to be processed in a text, transfer to a).

g) Order the rules depending on their score.

The following are some examples from error correction rules:

Ex1. rectify_segmentation_error (**concat**, 莫晨|N4|月|N, 1, N4, 前锋|N, 在|P)

Ex2. rectify_segmentation_error (**split**, 本周日和|N5, 1|3, R|T|C, 参加 |V, 阿曼|N7)

Ex3. rectify_segmentation_error (**slide**, 宏|G|远门|N|将|D, right1|right1, N|N, 使|P, 猝不及防|I)

Ex4. rectify_tag_error (赛|V, N, 小组|N, 时|N)

Here note that D, G, I, N, N4, N5, N7 and P represent an adverb, a morpheme, an idiom, a general noun, a PN, a LN, a transliterated PN or LN and a proposition respectively.

Considering the requirements of context constraints for different rules, we divide the rules into three rule types, that is, whole context sensitive, preceding context sensitive and whole context free, manually. Hence, this prevents new errors caused by using error correction rules. The algorithm applied to correct errors is given as follows:

a) Input a sentence by automatic word segmentation and POS tagging.

b) Retrieve the transformation rule library in such sequence: whole POS context constraints, preceding POS context constraints and without context constraints. If one of rules in rule library is matched, execute the corresponding transformation action.

c) Correction of word segmentation errors precedes correction of POS tagging errors.

d) If there is still sentence to be processed, transfer to a).

For example, the above rule for correcting segmentation error (Ex.3) is applied to the sentence with the corresponding errors, the rectified sentence is shown as follows:

使|P|宏远 |N|门将|N|猝不及防|I

3. NAMED ENTITY RECOGNITION

We make use of Finite-State Cascades (FSC) [1] as analysis mechanism for NER in our system. FSC is automatically constructed by the POS rule set with the semantic constraints. It consists of three levels. Each level has a NE recognizer, that is TN, CT and PI recognizer. In recognition, if two rules all are matched, we select maximum length match as final match.

The basic recognition procedure is described in a following example:

上海|N5|申花|N|队|N|在|P|百事可乐|N|甲|N| A |QT|联赛|N|中|F|击败|V| 对手|N|吉林|N5| 敖东|N|队|N|。|W|

Shanghai Shenhua Team defeated the opponent – Jilin Aodong Team in the Pepsi First A League Matches.

```
L3  -----------TN  P  -----------------CT  N  V   PI   -----------TN  W
L2  -----------TN  P  -----------------CT  N  V   N    -----------TN  W
L1  -----------TN  P  N   N QT N  N  V   N    -----------TN  W
L0  N5  N  N  P   N   N QT N  N  V   N    N5  N   N  W
    上海 申花 队 在 百事可乐甲 A 联赛 中 击败 对手 吉林 敖东 队 。
```

L_i is a level of FSC, which corresponds to a NE recognizer. That is, TN, CT and PI recognizer are located on L_1, L_2 and L_3 respectively. The shadow under the word or phrase means that it is a NE.

Sometimes there is no keywords combined with TN or CT. For such situation, domain verbs are collected and verb valency [7] is applied to analyze the constituents in sentences. Additionally, we use TN and CT context clues to determine whether the current entity is one of them.

4. EXPERIMENTAL RESULT

The Chinese Named Entity Recognition System (CNERS) has been implemented with Java 2 (ver.1.4.0) under Windows 2000. The recognized text can be entered from disk or directly downloaded from WWW. HowNet Knowledge Dictionary [4] is used to provide English and concept explanation of Chinese words in the recognized results. The system, which has been tested on a big corpus in the sports domain, is fast and reliable.

Moreover, 20 Web news about football sports from Jie Fang Daily (http://www.jfdaily.com/) in May 2002 have randomly been chosen and tested. Recognition results are compared with and without error correction. Average recall and precision are shown in Figure 2 and 3.

The experimental result has indicated that the performance for most of NEs in our system has been improved, the average recall and precision of six NEs are increased by 14%.

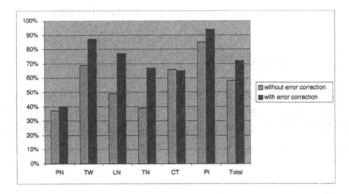

Figure -2. Recall Comparison

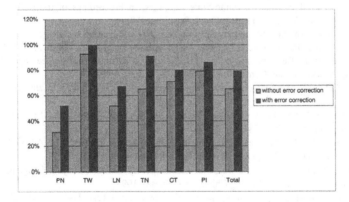

Figure -3. Precision Comparison

But the result has also revealed the recall and precision of PN are still lower than other NEs. The reason is that a Chinese name can be combined with nearly every Chinese character, so that the right boundary of a name is difficult to determine. Therefore, the rules from machine learning can not cover most of name's errors.

5. CONCLUSIONS

In Chinese IE investigation, we note that the errors from word segmentation and POS tagging have adversely affected the performance of NER to a certain extent. We utilize a machine learning technique to perform error correction for word segmentation and POS tagging in Chinese texts before NER is done and improve the recognition performance for most NEs. In addition, FSC is used as an analysis mechanism for Chinese NER, it is suitable and reliable. Such a hybrid approach used in our system synthesizes the advantages of knowledge engineering and machine learning.

ACKNOWLEDGEMENTS

This work is a part of the COLLATE (Computational Linguistics and Language Technology for Real World Applications) project under contract no. 01INA01B, which is being supported by the German Ministry for Education and Research.

REFERENCES

1. Abney S. Partial Parsing via Finite-State Cascades. In Proceedings of the ESSLLI '96 Robust Parsing Workshop. Prague, Czech Republic, 1996.
2. Brill E. Transformation-Based Error-Driven Learning and Natural Language Processing: A Case Study in Part of Speech Tagging. Computational Linguistics. Vol. 21, No. 4, 1995.
3. Chen H.H. et al. Description of the NTU System Used for MET2. Proceedings of 7th Message Understanding Conference, Fairfax, VA, U.S.A., 1998.
4. Dong Z.D. and Dong Q. HowNet. http://www.keenage.com/zhiwang/e_zhiwang.html, 2000.
5. Hockenmaier J. and Brew C. Error-Driven Learning of Chinese Word Segmentation. Communications of COLIPS 8 (1), 1998.
6. Kameyama M. Information Extraction across Linguistic Barriers. In AAAI Spring Symposium on Cross-Language Text and Speech Processing, 1997.
7. Lin X.G. et al. Dictionary of Verbs in Contemporary Chinese. Beijing Language and Culture University Press. Beijing China, (In Chinese), 1994.
8. Liu K.Y. Automatic Segmentation and Tagging for Chinese Text. The Commercial Press. Beijing, China. (In Chinese), 2000.
9. Palmer D. and Burger J. Chinese Word Segmentation and Information Retrieval. In AAAI Spring Symposium on Cross-Language Text and Speech Retrieval, Electronic Working Notes, 1997.
10. Wilks Y. Information Extraction as a Core Language Technology. In Maria Teresa Pazienza editor, Information Extraction: A Multidisciplinary Approach to an Emerging Information Technology, LNAI 1299, pages 1-9. Springer, 1997.

IMPLEMENTATION OF VISIBLE SIGNATURE*

Kefei CHEN, Zheng HUANG, Yanfei ZHENG
Department of computer Science and eigineering
Shanghai Jiaotong University
Shanghai 200030, PR CHINA
{chen-kf,huang-zheng,zheng-yf}@cs.sjtu.edu.cn

Abstract   This paper introduces a novel electronic document (like word, excel file format) authentication scheme. In this scheme, user's digital signature string is embedded into user's rubber stamp image using digital watermark technology. We call the watermarked image visible signature and call the authentication scheme visible signature scheme. Combined with the digital watermark technology, PKI technology and smart card technology, the scheme is very secure for authentication, is easy to use and can works seamlessly with popular applications like word, excel and other electronic document processors.

Keywords: digital signature, digital watermark, visible signature scheme

1.   Introduction

There are lots of electronic documents flowing around the Internet and Intranet everyday. On receiving an electronic document, how can you determine that the document is originated from the man who declares to write it and determine that the document has not been altered since it was written? When you send an electronic document to the other people, how can you confirm them that the document is really written by you? Of course, these problems can be solved using the PKI technology. But it is not a practical way to use PKI technology directly [2], since you must at least attach a signature string to the document. The purpose of this paper is to find a more straightforward method based on PKI technology for the purpose of authentication of electronic documents. The straightforward method is much like that in the traditional way

*This work was partial supported by NSFC under the grant 69973031 and 90104005.

G. Hommel and S. Huanye (eds.), The Internet Challenge: Technology and Applications, 37–44.
© 2002 Kluwer Academic Publishers.

38

of authentication of document on paper format, that is you can write you signature on the paper or put your rubber stamp on the printed document to authenticate the document. Our method is to embedded your visible signature into the electronic document, so the embedded visible signature could be used as the visible authentication to your electronic document.

1.1   Basic Concept of Visible Signature

The visible signature can be described using Figure1.

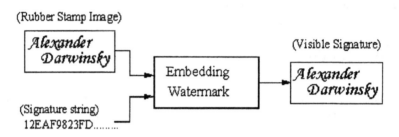

Figure 1.   Concept of Visible Signature

From Fig 1, we could see that the outlook of visible signature seems just like someone's rubber stamp image that is very familiar with many users. The basic idea of visible signature is embedding the digital signature string in to user's rubber stamp image using digital watermarking technology. We call the watermarked rubber stamp image visible signature. Although the watermarking technology may introduce slightly deformation to the rubber stamp image, everyone could recognize the visible signature as his/her rubber stamp image at a glance. If we embed the visible signature into a formatted document, the digital signature string that embedded into visible signature could be used for authentication purpose like that in digital signature scheme.

1.2   Features of Visible Signature

Visible signature has the following advantages compared with simply use digital signature:

- Easy to use. Most users are familiar with putting the hand writing signature or rubber stamp image on a printed document to authenticate the document. When comes to the electronic document, most of them are confused by the abstract digital signature concept to authenticate an electronic document. With visible signature, it will become easier for them to understand, thus they will accept the concept more easily.

- Easy to integrated with existing applications. Most word processing applications, like Word, Excel and PDF, support inserting a image into the document. So the visible signature could be easily integrated with these applications.

1.3 Electronic Rubber Stamp

The traditional rubberstamp is used to produce a rubberstamp image on the printed document. When comes to visible signature, the rubberstamp becomes a hardware that is used to produce the visible signature. The hardware is response for storing the original rubber stamp image of the user, producing digital signature string for the document and making visible signature. Using the hardware as a rubberstamp, the security of the visible signature system can be guaranteed. We call the hardware electronic rubber stamp.

1.4 Visible Signature Application

Visible signature application is a program that runs on user's computer. It is the interface of the visible signature scheme to the user. Visible signature application works together with the word processes (like MS word, Excel) and the electronic rubber stamp. The user uses visible signature application to visibly sign an electronic document and check if a visible signature is valid.

1.5 Outline of this Paper

This paper is organized as follows: In section 2, we will give an overview of the visible signature scheme. The digital watermark technology that used in this scheme is described in section 3. Section 4 will describe electronic rubberstamp. We will give a sample application of visible signature in Section 5 and we will round up this paper in section 6.

2. Overview of this Scheme

Like normal digital signature scheme, the visible signature scheme has two basic functions that are signing and checking. Fig 2 and Fig 3 give block diagrams of the two basic functions. In order to simplify the explanation, we do not deal with how the user gets the certificate from a CA.

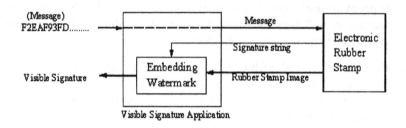

Figure 2. Block Diagram of Signing

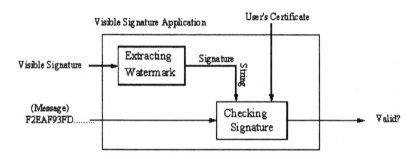

Figure 3. Block Diagram of Checking

3. Digital Watermark

We use digital watermark technical to embed the user's digital signature string into the rubber stamp image. Digital watermark technology has received many researchers' attention since it plays an important role in information hiding and copyright protection [4]. There are many digital watermark technologies in the literature as well. The digital watermark technology we need must have the following properties:

- Accuracy: In the checking phase, we must retrieve the user's digital signature string from the visible signature. The retrieved string must be 100% accuracy, which means every bit of the signature string must be retrieved correctly.

- Robustness: The signature image will be placed into the electronic document. In order to reduce the document size, many word processors, like MS Word, use some public or propriety algorithms to compress the images in the document. The digital watermark technology must stand with these types of compression.

- Visibility: The requirement for the visibility of the digital watermark technology is not very strict. We just need the embedding process doesn't introduce severely damages to the rubber stamp image.

- Oblivious: A watermark technology is oblivious means the original image is not needed when retrieving the embedded information. A signed document may be checked by a number of other users. It is not practical to make the other users have the rubberstamp image of the user who signed the document.

- Capability: A typical signature string, like RSA signature, has 1024 bits. A typical rubberstamp image is less than 200×200 pixels. The digital watermark technology must embed at least 1024 bits in such an rubberstamp image.

3.1 Implementation of Digital Watermark Technology

Based on the requirements list above, we implement our digital watermark technology as follows. In order to make the digital watermark to be an oblivious watermark, we fix the rubberstamp image to be a 200×200 24bits image. This image has separated RGB channel. Notice that rubberstamp images in real life are mono-color image and the mono-color are always red. So we fix the background of the rubberstamp image to be value that RGB channels are all 170. So the background is light gray color. For the strokes of the rubberstamp image, we use the color with red value 255, green value 170 and blue value 170. This color is an almost red color. From the knowledge of human visual system, we could see that human eyes are not very sensitive to the blue channel. Embedding watermark into the blue channel of the rubberstamp image could introduce less unpleasing results. If we only embed watermark into the blue channel, the watermark is an oblivious watermark. This is because of the reason that the blue channel value of every pixel in the rubber stamp image is 170.

To the robustness of the watermark, we use the famous robust digital watermark scheme that presented in [1]. Here we recapitulate the embedding method. The original rubber stamp image is divided into 8×8 blocks. DCT transform is performed on each 8×8 block. In order to make the watermark robust enough, we choose two low frequency coefficients (1,2) and (2,1) of the 8×8 DCT coefficients to embed the watermark. One bit is embedded into each selected coefficient. Let the value of the selected coefficient is a. If the embedded bit is 1, the value of the coefficient is set to $a + 0.1$. If the embedded bit is 0, the value of the coefficient is set to $a - 0.1$. After embedding all bits, performance the IDCT transform to get the watermarked image: the visible signature. The process of retrieve is straightforward. The experiments we did show that this watermarking method is robust to the image compression algorithm used in MS Word and Excel processor. We could get the embedded signature string from the visible signature image that placed in the word document with 100% accuracy.

The rubber stamp image is 200×200 bitmap image with 24bit colors. We only use the blue channel of the image to embed watermark. There are 625 8×8 blocks in the rubber stamp image with 2 bits embedded into each block. So the capacity of the watermark is $625 \times 2 = 1250$ bits. A typical RSA signature is 1024 bits. So there is plenty of room to hide a RSA signature and some auxiliary data into the watermark.

4.   Electronic Rubber Stamp

The electronic rubber stamp is a hardware that acts as a rubber stamp. The difference is that the traditional rubber stamp works on paper while the electronic rubber stamp works with computer.

Electronic rubber stamp is an USB device with a smart card, a smart card reader and some flash memory integrated in it. It provides not only memory capacity, but computational capability as well. The security of the electronic rubber stamp is based on the security of smart card [3]. A PIN number is needed when accessing the electronic rubber stamp. In the visible signature scheme, the electronic rubber stamp provides functions listed below:

- Providing storage space: There is 32K flash memory in the electronic rubber stamp. The memory is used to store some personal information of the electronic rubber stamp. The most important among them are:

    - Private key of the electronic rubber stamp
    - Certificate of the electronic rubber stamp
    - Self-signed certificate of the CA and certificate chain
    - The rubber stamp image (in compressed format)
    - Log of signature records

- Providing cryptography functions:

    - RSA(1024bit) sign/verify and encryption/decryption
    - RSA(1024bit) keys (private key and public key) generation

- SHA-1 Hash function
- Random number generation using hardware

- Providing signature log functions:

  - Recording the name of the signed document and the time when signing
  - Retrieving all the records

5. Sample Application

We have built a visible signature application that integrated with MS word processor. This application is represented as a toolbar in MS word environment. Fig 4 shows the interface of visible signature application (toolbar) and a visible signature that placed in MS word document. We call the application ENOD.

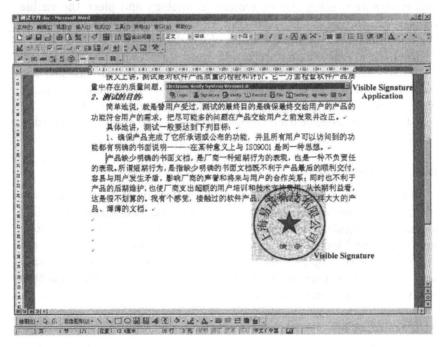

Figure 4. Sample Application

ENOD mainly provides user the following functions:

- Login: Let user enter the PIN number of the electronic rubber stamp to access the electronic rubber stamp.

44

- Sign: Let user place his visible signature at any place in the document when he is sure that he wants to sign the document.

- Check: Let user check whether the visible signature in the current document is a valid signature from the person who signed it.

- View Log: Let user view the logs in the electronic rubber stamp. The user could know when and on which document his electronic rubber stamp has been used.

To sign a MS word document, ENOD first extracts the content of the MS word document. The content includes all the words, all the pictures (not including the visible signature), all the tables and all the format information in the document. Then, ENOD sends the hash value of the content to electronic rubber stamp to get the signature and the original rubber stamp image. After the step above, ENOD could make the user's visible signature using the watermark technology and place the visible signature in the document. Note that what the user signed is not directly the document itself, but the content of the document.

To check a visible signature in MS word document, ENOD first extracts the content of the MS word document as that in sign process. Then ENOD extracts the signature string in the visible signature using watermark technology. After doing these, ENOD gets the user's certificate from local cache or from online OSPF server to check the validation of the signature string.

6.   Conclusions

In this paper, we introduce a new conception "visible signature", explain the implementation and give our sample application ENOD. Visible signature integrates the digital watermark technology, PKI technology and smart card technology together and has the feature of easy to use and easy to accept by user. It is the electronic form of the rubber stamp that users are all familiar with. We believe that the visible signature will bring a new scene in PKI applications.

References

[1] I. Cox, J. Kilian, T. Leighton, T. Shamoon: Secure Spread Spectrum Watermarking for Multimedia, IEEE Trans. on Image Proc., Vol.6, No.12, 1997.
[2] The Open Group: Architecture for Public-Key infrastructure, Available at http://www.opengroup.org/publications/catalog/g801.htm
[3] CHAN, Siu-cheung Charles: An Overview of Smart Card Security, Available at http://home.hkstar.com/ alanchan/papers/smartCardSecurity/
[4] G. Voyatzis, N. Nikolaidis, I. Pitas: Digital Watermarking: An Overview, 9th European Signal Processing Conference (EUSIPCO'98)

INTERNET-BASED ROBOTIC WORKCELL CALIBRATION & MAINTENANCE

Lu Tiansheng   Guo Jianying
School of Mechanical Engineering, Shanghai Jiao Tong University (200030)
E_mail: tslu@mail.sjtu.edu.cn, gjy_dlut@hotmail.com

Abstract: In this paper, a method using the active vision for the robot kinematic calibration and workpiece automatic localization is proposed. By the identified robot kinematic parameters, the error compensation procedure is carried out, resulting in the robot positioning accuracy improvement. With the workpiece localization, the robot program can be revise to drive the remote robot to work on the exact point with the off-line programming pose.

Key words: Internet-based robotic work-cell, Robot calibration, Workpiece localization

1.   INTRODUCTION

Nowadays, manufacturing companies have been plagued by depletion of natural resources, diversified customer requirements, reduced product life cycles and increased production complexity. Internet that constitutes of ubiquitous network and the technological development make this problem settled and arising tele-manufacturing and Internet-based robotic workcell.

The first generation of Internet robotic system came into existence in 1994. The Mercury project was first carried out at that time to build a six-DOF tele-manipulator, allowing users to pick up and manipulate various objects within its reach. The tele-operation system at the University of Western Australia can pick and put down blocks, the tele-manufacturing work-cell is set up in National University of Singapore.

All these are mainly based on robotic arms directly controlled by human operators with the visual feedback from the field. These systems can be used

G. Hommel and S. Huanye (eds.), The Internet Challenge: Technology and Applications, 45–53.
© *2002 Kluwer Academic Publishers.*

for fun or demonstration of the telecommunication technique used in the robotics and manufacturing. In this paper, the schematic of a new generation robot work-cell is proposed.

2.   INTERNET-BASED ROBOTIC WORKCELL

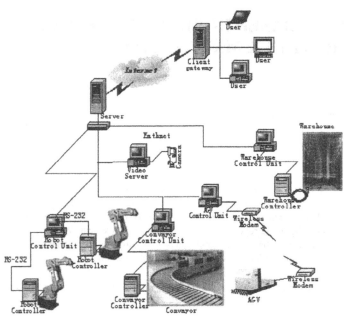

Figure 1 Outline of the tele-manufacturing robotic work-cell system

As shown in Fig.1, the tele-manufacturing robotic workcell consists of robot manipulators, conveyor, AGV, over head camera-video system and warehouse. The control unit associated to the robot manipulators, conveyor, warehouse and the AGV is connected to the central control workstation. The control units associated to robot manipulators are connected to robot controllers through the RS-232. AGV communicates with its control unit through the wireless modem. The overhead camera can be controlled to get a panoramic view, which makes it possible to locate the trouble.

3.   ROBOTIC WORKCELL CALIBRATION

This generation of Internet-based robotic workcells use off-line programming technique to generate the robot work program on the remote client workstation with the off-line programming software.

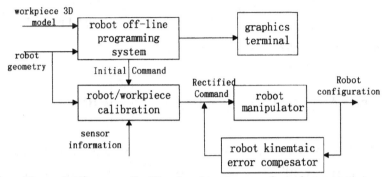

Figure 2 Diagram of calibration for tele-manufacturing system

As shown in Fig.2, the robotic work-cell calibration for the Internet-based robotic workcell includes the robot calibration and the work-piece relative pose calibration, namely work-piece localization. The robot program generated by the off-line programming module will be transferred to the server first via the Internet. The server then transfers the program to the robot/work-piece calibration module. In this calibration module, the transformation between the programming reference frame attached to the work-piece and the robot base coordinate system is calibrated. After the work-piece relative pose calibration, the rectified robot program can be used to drive the robot to work properly.

In the following section, the coplanar pattern based camera calibration for robot pose measurement is introduced. By the robot-vision system, the work-piece localization can be fulfilled automatically.

3.1   Coplanar rigs for camera calibration

Let **P** be a point in space with the coordinate vector $\mathbf{XX} = [X; Y; Z]$ relative to the grid reference frame and $\mathbf{XX}_c = [X_c; Y_c; Z_c]$ is the coordinate vector of **P** in the camera reference frame. The relationship between **XX** and $\mathbf{XX}_c$ satisfies: $\mathbf{XX}_c = \mathbf{R} * \mathbf{XX} + \mathbf{t}$, where $\mathbf{R} = \begin{bmatrix} \mathbf{r}_1 & \mathbf{r}_2 & \mathbf{r}_3 \end{bmatrix}^T$ is the rotation matrix and **t** is translation vector.

3.1.1   Computation of Homographies

Here the coplanar pattern is used for camera calibration. Without loss of generality, we assume the model plane is on $Z = 0$ of the world coordinate system. Let's denote the ith column of the rotation matrix **R** by $\mathbf{r}_i$.

Denoting $\tilde{\mathbf{m}}_p = \begin{bmatrix} x_p & y_p & 1 \end{bmatrix}^T$ and $\tilde{\mathbf{M}} = \begin{bmatrix} X & Y & 1 \end{bmatrix}^T$, a model point M and its image m_p is related by a homography $\mathbf{H}$:

$$s\tilde{\mathbf{m}}_p = \mathbf{A}\begin{bmatrix} \mathbf{r}_1 & \mathbf{r}_2 & \mathbf{r}_3 & \mathbf{t} \end{bmatrix}\begin{bmatrix} X \\ Y \\ 0 \\ 1 \end{bmatrix} = \mathbf{A}\begin{bmatrix} \mathbf{r}_1 & \mathbf{r}_2 & \mathbf{t} \end{bmatrix}\begin{bmatrix} X \\ Y \\ 1 \end{bmatrix} = \mathbf{H}\tilde{\mathbf{M}} \qquad (1)$$

By abuse of notation, we still use M to denote a point on the model plane, but $\mathbf{M} = \begin{bmatrix} X, Y \end{bmatrix}^T$. Thus, we have $\tilde{\mathbf{M}} = \begin{bmatrix} X & Y & 1 \end{bmatrix}^T$. A model point $\tilde{\mathbf{M}}$ in the grid reference frame and its image $\tilde{\mathbf{m}}_p$ can be related by a homography: $s\tilde{\mathbf{m}}_p = \mathbf{H}\tilde{\mathbf{M}}$. Where s is a scalar factor and $\mathbf{H} = \mathbf{A}\begin{bmatrix} \mathbf{r}_1 & \mathbf{r}_2 & \mathbf{t} \end{bmatrix}$ is a 3x3 matrix. h_1, h_2, h_3 are the 1st, 2nd and 3rd column of $\mathbf{H}$. let $\mathbf{x} = \begin{bmatrix} \mathbf{h}_1^T, \mathbf{h}_2^T, \mathbf{h}_3^T \end{bmatrix}^T$, we have $\begin{bmatrix} \tilde{\mathbf{M}}^T & \mathbf{0}^T & -x_p\tilde{\mathbf{M}}^T \\ \mathbf{0}^T & \tilde{\mathbf{M}}^T & -y_p\tilde{\mathbf{M}}^T \end{bmatrix}\mathbf{x} = 0$. When we are given n points, we have n above equations, which can be written in matrix equation as $\mathbf{Lx} = \mathbf{0}$. As $\mathbf{x}$ is defined up to a scale factor, the solution is well known to be the eigenvector of $\mathbf{L}^T\mathbf{L}$ associated with the smallest eigenvalue. Rearranging the elements of the vector $\mathbf{x}$, we get the 3×3 matrix $\mathbf{H}$.

Zhang first compute the matrix $\mathbf{B} = \mathbf{A}^{-T}\mathbf{A}^{-1}$, then computing the matrix $\mathbf{A}$, $\mathbf{R}$ and vector $\mathbf{t}$ [5]. In this paper, camera parameters is computed though the orthogonality of the vanishing points and the intrinsic characteristic of the homography.

3.1.2   Parameters estimation using the orthogonality of the vanishing point and the intrinsic characteristic of the homography

In this stage, only the focal lengths f_x and f_y are estimated while assuming that the principal point locate at the center of the image and the skew coefficient τ equals 0. In affine space, vanishing point is a point at which receding parallel lines seem to meet when represented in linear perspective. The two vanishing points associated to the lines orthogonal to each other are also orthogonal in the dual space.[2] That provides scalar constraint for the focal lengths f_x、 f_y:

$$\frac{a_1 a_2}{f_x^2} + \frac{b_1 b_2}{f_y^2} + c_1 c_2 = 0, \quad \frac{a_3 a_4}{f_x^2} + \frac{b_3 b_4}{f_y^2} + c_3 c_4 = 0 \qquad (2,3)$$

Where a_i, b_i and $c_i (i = 1, 2, 3, 4)$ are the known pixel coordinates. Through Equ.2 and 3, it is easy to find the initial value for f_x、f_y.

From Equ.1, it is easy for us to get $s\mathbf{A}^{-1}\tilde{\mathbf{m}} = [\mathbf{R} \ \ \mathbf{t}]\tilde{\mathbf{M}}$, that is, $s\tilde{\mathbf{m}}_n = \mathbf{H}'\tilde{\mathbf{M}}$, where $\mathbf{H}' = [\mathbf{R} \ \ \mathbf{t}] = [\mathbf{r}_1 \ \ \mathbf{r}_2 \ \ \mathbf{t}]$ and $\mathbf{m}_n$ is the normalized (pinhole) image projection. Using the method for the computation of the homograph introduced above, we can get $\mathbf{H}'$, the 1st and 2nd columns are $\mathbf{r}_1$ and $\mathbf{r}_2$ column vector of the rotation matrix respectively and the 3rd column is the translation vector. By the orthogonality of the rotation matrix we can get $\mathbf{r}_3$ through $\mathbf{r}_3 = \mathbf{r}_1 \times \mathbf{r}_2$.

3.1.3   Camera intrinsic & extrinsic parameters optimization

Let $p_i (i = 1, ..., N)$ be the observed image projections of the rig points P_i and let $\bar{\mathbf{p}}_i = \begin{bmatrix} p_x^i & p_y^i \end{bmatrix}^T$ be their respective pixel coordinates. Points p_i are detected using the standard Harris corner finder. The estimation process consists of finding the set of calibration unknowns that minimizes the re-projection error. Therefore, the solution to that problem may be written as follows:

$$\{f_x, f_y, c_x, c_y, \mathbf{k}_c, \mathbf{R}_c, \mathbf{T}_c\} = \min \sum_{i=1}^{N} \left\| \bar{\mathbf{p}}_i - \mathbf{\Pi}\left(\mathbf{R}_c \bar{\mathbf{X}}_c^i + \mathbf{T}_c\right) \right\|^2 \qquad (4)$$

Where $\mathbf{\Pi}(\bullet)$ is the image projection operator, that is, with the operation of the operator we can get the pixel coordinate vector $\mathbf{x}_p$ in the image plane of the point p, $\|\bullet\|$ is the standard distance norm in pixel unit. With the initial value estimation of parameters, this non-linear optimization problem may be solved using standard gradient descent techniques.

3.2   Robot hand-eye calibration with GA

Robot hand-eye calibration is the computation of the relative position and orientation between the robot gripper and a camera mounted rigidly on the robot end-effector. The hand-eye relationship are first formulated by Shiu and Ahmad[3]: $\mathbf{AX} = \mathbf{XB}$. The matrix $\mathbf{A}$ and $\mathbf{B}$ denote the transformations between the first and the second position of the robot hand(in the robot coordinate system) and the camera(in the camera system) respectively and **X** denotes the transformation between the hand coordinate system and the camera coordinate system, i.e. the hand-eye calibration.

In most cases, the hand-eye calibration can be simplified by using the possibility to move the robot in a controlled manner and investigating the arising motion field of points in the images. In this paper, we formulate the minimization problem as $f = \min_i \sum_i \left\| \mathbf{A}_i \mathbf{X} - \mathbf{X} \mathbf{B}_i \right\|_F$. By minimizing the residue, we can find a hand-eye transformation robust to the noise and disturbance existing in the process. GA optimization method is applied to solve the transformation **X** without the estimation of the initial value.

3.3   Pose measurement and parameters identification

By camera calibration we have the extrinsic parameters of the camera **A** at every position relative to the world coordinate system. By hand-eye calibration we can get the transformation **X**. Then robot end-effector pose relative to the world coordinate system satisfy: $\mathbf{T}_N = \mathbf{XA}$

The pose of the robot end-effector can be described as a function of its kinematic parameters. Due to robot parameter errors, the actual robot poses always deviates from those predicted by its controller. By the differential movement, we have $X_i' - X_i = J_i \Delta P, X_j' - X_j = J_j \Delta P$, where, X_i' and X_j' are the actual values of X_i and X_j . ΔP is the kinematic parameter deviation set, J_i and J_j are the corresponding identification Jacobian matrices. We can find $\left\| X_i' - X_j' \right\|^2 - \left\| X_i - X_j \right\|^2 = 2(X_i - X_j)^T (J_i - J_j) \Delta P$, where, $\left\| X_i' - X_j' \right\|$ is the true distance between two targets determined through the camera calibration and $\left\| X_i - X_j \right\|$ is the distance calculated by the robot system based on nominal parameter values.[1]

3.4   Model based robot position error compensation

We apply the method inverse model control procedure to compensate the kinematic parameter error. Say that the goal pose of the robot is Location N , the corresponding joint variable vector $\mathbf{q}_N$ by solving an inverse kinematic using the nominal kinematic parameter vector $\mathbf{\rho}_N$. Due to errors in the robot geometry, the robot will not reach N if the control command $\mathbf{q}_N$ is used.

Pose $\mathbf{T}_A$ of the robot under this control command can be estimated by solving a forward kinematic using the identified kinematic parameter vector $\mathbf{p}_A$. Let $d\mathbf{T}$ denote the difference between $\mathbf{T}_N$ and $\mathbf{T}_A$. Solving another

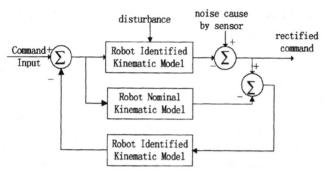

Figure 3   Inverse model based error compensation

inverse kinematic with $\mathbf{p}_A$, one obtains the $\Delta\mathbf{q}$, which can be combined with $\mathbf{q}_N$ to control the robot to move to location N.

3.5   Workpiece automatic localization

Conventionally, the work-piece was fixed by jigs, fixtures and hard gauges or setup manually. In both cases, it is expensive, and to the latter, it also needs works with high expertise. If the workpiece is fixed and positioned on the appropriate fixture with regular contours and transferred by the conveyor, it is possible for us to find the coordinate frame fixed on the pallet relative to the robot base coordinate frame. Thereafter we can find the position and orientation of the workpiece in the robot coordinate system. Otherwise, if the workpiece is arbitrarily placed and fixed but not well positioned on the general-purpose fixture, it is necessary to find the transformation between the point on the workpiece and the corresponding point on the CAD model used for the off-line programming. This method is called the workpiece localization and the way to position workpiece is called computer aided setup [4].

During the off-line programming stage, it is impossible for us to know the exact relationship between the robot base and the work-piece reference frame. And in the real manufacturing process, it is also difficult for us to guarantee all the work-piece with the same poses relative to the robot base when arriving at the processing position. Having the robot and hand-eye calibrated, the camera relative to the robot base coordinate system can be precisely determined. With the camera calibration, the transformation between the grid coordinate system and the camera reference frame also can be recovered with satisfactory accuracy. In this way, the frame associated to the work-piece relative to the robot base can be calibrated.

4.   INTERNET-BASED ROBOTIC WORKCELL MAINTENANCE

In the server applications, there is a watchdog demon to monitor all control units. The watchdog reads the status of each subsystem from the corresponding control unit, and control units collect the status and the error/warning message from its controlled devices. It is a hierarchy structure. The behavior and performance information of each device can be assessed and evaluated from the field with the watchdog demon. When the work-cell broken down, the control unit can find the error first and generate the BROKEN_DOWN messages and these messages then collected by the watchdog application. The watchdog analyzes and assesses these error messages, trying to locate the trouble. It sends another request to the assuming collapsed device to assure the initial estimation. If it was assured, then the watchdog demon sends the information to the user. With the error message, the user can control the overhead camera to locate the broken down parts and find the problem and give the solution in real time.

5.   CONCLUSIONS

In this paper, the first generation of tele-robotics system is reviewed and the shortcoming is given and the schematic of the robotic workcell introduced. To make the system to work perfectly, method of active vision based robot calibration and the workpiece localization is introduced, which makes the system flexible. The self-calibration and tele-maintenance makes the Internet-based robotic workcell feasible.

REFERENCE:

1. Chunhe Gong, Jingxia Yuan, Jun Ni. A self-calibration method for robotic measurement system. Transaction of the ASME. Journal of Manufacturing Science and Engineering, Vol.122, No.1, Feb.,2000, pp. 174-181
2. Jean-Yves Bouguet. Visual methods for three-dimensional modeling, PhD. thesis, California Institute of Technology , 1999
3. Y. C. Shiu and S. Ahmad. Calibration of wrist-mounted robotic sensors by solving homogeneous transform equations of the form AX=XB, IEEE Transaction of Robotics and Automation, Vol.5, No. 1, pp.16-27, 1989

4. Zexiang Li, etc. Geometric Algorithm for Workpiece Localization. IEEE Transactions on Robotics and Automation, Vol.14, No.6, Dec. 1998, pp.864-878
5. Zhenyou Zhang. Flexible Camera Calibration by Viewing a Plane from Unknown Orientation. the Proceedings of the Seventh IEEE International Conference on Computer Vision(1999),Vol.1,pp.667~673

TOWARDS MULTI-MODAL MOBILE TELE-PRESENCE AND TELEMANIPULATION

Christian Reinicke   Martin Buss

Control Systems Group, Faculty EECS, Technical University Berlin
Einsteinufer 17, D-10587 Berlin, Germany
reinicke@rs.tu-berlin.de, M.Buss@ieee.org

Abstract   This paper presents a *mobile* multi-modal telepresence and teleaction (MTPTA) system for telemanipulation in wide area real remote environments. An interactive stereo visual telepresence system allows human operators to actively be present in the remote environment with high fidelity 3D visual immersion. The mobile teleoperator is remotely controlled by IP communication and commonly available WWW standard software.

Keywords: Multi-modal telepresence, telemanipulation, WWW robot control.

Introduction

Multi-modal *mobile* telepresence and teleaction systems include classical teleoperation and telemanipulation systems extended by mobility functions of a mobile robot. An important advantage of the combination of telepresence and teleaction with a mobile robot, is that it allows the human operator to perform actively in wide area remote environments. One of the central issues in modern telepresence systems is multi-modality in the human system interface (HSI) accompanied by appropriate sensing techniques at the teleoperator site comprising theoretically all the human senses. In current technical applications most important and only partly realized are the visual, auditory, and haptic senses, see e.g. Baier et al., 2000; Burdea and Zhuang, 1991; Buss and Schmidt, 1999; Buss and Schmidt, 2000; Fischer et al., 1996; Kaczmarek et al., 1994; Lee, 1993 and Maneewarn and Hannaford, 1998.

The application areas of *mobile* telepresence and teleaction systems are countless, to name only a few: tele-installation, tele-diagnosis, tele-service, tele-maintenance, tele-assembly, tele-manufacturing, miniature

G. Hommel and S. Huanye (eds.), The Internet Challenge: Technology and Applications, 55–62.
© *2002 Kluwer Academic Publishers.*

or micro mechatronics, inner and outer space operations, tele-teaching, tele-medicine, tele-surgery, tele-shopping, etc.

Recently, systems with multi-modal human interaction are of increasing interest to the research community. One of the most important enabling factors for this development are immense advances in multimedia and Internet technology. In cooperation with the Institute of Automatic Control Engineering of the Technische Universität München we developed a MTPTA system introduced in Section 2 with active stereo visual telepresence capability. The multimodal HSI is located in Munich and the *mobile* teleoperator in Berlin.

This paper mainly discusses the *mobile* teleoperator system and WWW remote mobility control. Without the need of special hardware for the HSI, commonly available WWW standard software is used. Here we present a remote environment map visualization using SVG/XML and standard WebCam-browser based visual telepresence.

1. Mobile Multi-Modal Telepresence Systems

The *general structure of a mobile multi-modal telepresence and teleaction (MTPTA) system* is depicted in Figure 1. On the operator-site the human operator gives multi-modal command inputs to the HSI using motion, voice, or symbolic input devices. The commands are transmitted to the executing teleoperator on the remote-site across barriers (e.g. communication or scale). The teleoperator — usually an executing robotic system such as a mobile service robot — is controlled according to the commands received from the human operator. Sensors mounted on the teleoperator gather data about the interaction between the teleoperator and the environment. Typically visual, acoustic, force, and

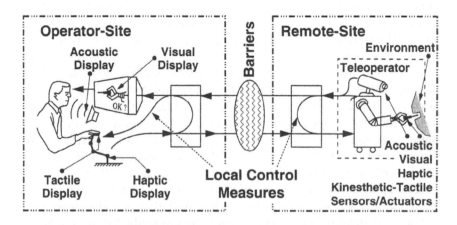

Figure 1. Multi-modal telepresence and teleaction system (TPTA).

tactile sensors are employed. Measured data is transmitted back to the human operator and displayed using modality dependent hardware in the HSI; the term *multi-modal display* includes the conventional terms of *graphic display* and *haptic display*.

One of the key issues in MTPTA system design and operation is the degree of coupling between the human operator and the teleoperator. If the operator gives symbolic commands to the teleoperator by pushing buttons and watching the resulting action in the remote environment the coupling is *weak*. The coupling is comparably *strong* for the kinesthetic modality in a bilateral teleoperation scenario. Commonly, the motion (force) of the human operator is measured, communicated, and used as the set-point for the teleoperator motion (force) controller. On the other hand the resulting forces (motion) of the teleoperator in the remote environment are sensed, communicated, and fed back to the human operator through the HSI. The literature distinguishes between *remote, shared, cooperative, assisting, semi-autonomous, symbolic,* and *trading control,* see e.g. Anderson, 1996; Buss, 1994; Buss and Hashimoto, 1993; Buss and Schmidt, 1999; Fischer et al., 1996 and Sheridan, 1992 for a more detailed discussion.

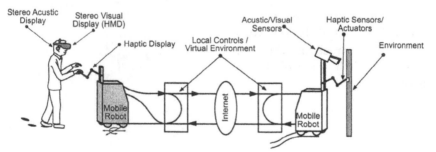

Figure 2.   Mobile telemanipulation system (MTMS) structure.

2. Mobile Telemanipulation System

The principle construction of a *mobile* telepresence and telemanipulator system (MTMS) is shown in Figure 2. The MTMS is a MTPTA system, where the HSI and the teleoperator are not fixed at a specific location, but are both mobile in a relatively wide area.

2.1 System Architecture

The mobile omnidirectional robot used for the mobile teleoperator platform was first developed by Hanebeck and Saldic, 1999 and provided to the Technical University Berlin by Sedlbauer Inc. This platform has an onboard integrated Single-Board-Computer for control, odometry, etc. Way points, desired velocities or trajectories are commanded via a CORBA-interface. On the user HSI side the x- and y-position from a 6 degree-of-freedom (DOF) "Flock of Birds" sensor are taken and sent to the mobile teleoperator system shown in Figure 3. On top of the mobile platform a laser range sensor, control and communication PCs, and the stereo visual system and head controller can be seen.

The audio-visual-head, see Figure 4(a), on the mobile robot is based on the system from Baier et al., 2000. It is constructed with two cameras with variable focal length and motor adjustable vergence and two microphones, both are mounted on a 3 DOF pan-tilt-roll head. The two video signal streams are synchronized by hardware. On the user side the image-streams are displayed on a "5DT HMD 800" head mounted display (HMD) Figure 4(b). For human interaction the 6 DOF sensor "Flock of Birds" is fixed on the HMD to track the three angles (roll, pitch, yaw) of the human head.

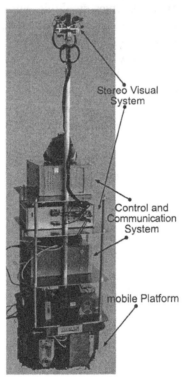

Figure 3. Picture of the MTMS.

2.2 Active Stereo Visual Telepresence

The goal of any telepresence system is to achieve high-fidelity multimodal human immersion with time consistent information display, low bitrate, and low time-delay as important factors to consider for a MTPTA system. Therefore several problems have to be solved including IP communication restrictions, synchronization of the left and right image transmission, and computation time reduction.

Hence the stereo visual telepresence system is an improved version of the one developed in the Institute of Automatic Control Engineering, Technische Universität München, see Baier et al., 2000; Baier et al.,

2002. This visual system has a minimum of network communication, and automatically synchronizes the two video data streams.

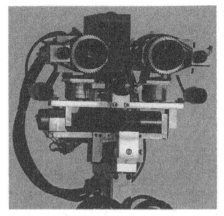 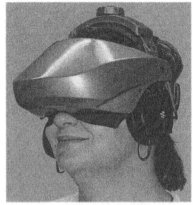

(a) Camera head   (b) Head Mounted Display

Figure 4.   Camera head and HMD of the active stereo telepresence system.

3. WWW-Control of Telerobots

For a simple and generically applicable remote control method for mobile teleoperators, it is beneficial to use standard WWW technology because there is no need for dedicated special hardware.

3.1 Position Visualization Using SVG/XML

For the visualization of the robot a SVG-map (scalable vector graphics) of the remote environment is used. This format is an open graphic standard, it is XML-conform and platform independent. XML is always a plain ASCII-format, so that it is simple to make changes in the data. Additionally, SVG is a vector format, so there are no quality problems by scaling the graphics. Most CAD software can also export SVG format. Another very important feature is event handling in the SVG specification. The event handling makes it relatively easy to update the position of the robot in the map. The robot or a communication-task can send events to the SVG-engine during motion.

In Figure 5 the visualization of the robot is shown. The display has three parts: the map of the institute with the robot (square), the numerical coordinates of the robot, and the control interface lower right side. With the arrow keys the user can send commands to the robot. The mobile platform will execute the moves, if it is possible and there is no obstacle (wall, closed door, or other object). During movement the robot position is periodically updated on the map.

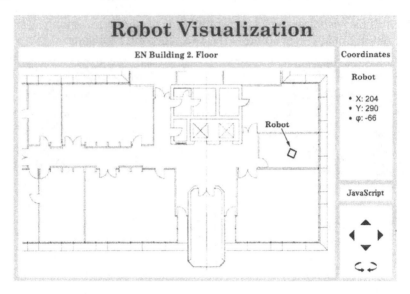

Figure 5. Visualization of the robot position.

3.2 WebCam-Browser Visual Telepresence

For users without HMD hardware to have visual feedback, there is an additional USB-WebCam mounted on the mobile teleoperator system.

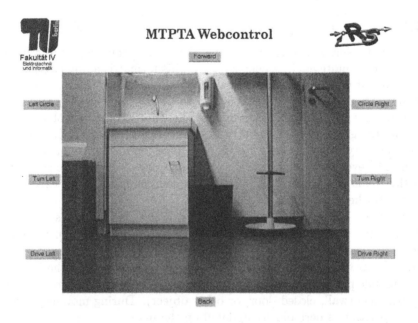

Figure 6. WWW control interface for the MTPTA.

The WebCam images can easily be made available to browser clients by a http-server. Figure 6 shows the WWW control interface for the mobile teleoperator system. It displays the live *image* stream of the WebCam in the middle and around the image are control buttons. The control part of the interface goes directly to the CORBA-Interface of the mobile teleoperator by sending the commands from the pressed buttons to the robot. So one can control and observe the movement of the teleoperator through this browser interface.

4. Conclusions and Future Work

An overview of the general structure of a *mobile* multi-modal telepresence and teleaction (MTPTA) system has been given. The MTPTA system may be applied with benefits in a wide range of industrial, commercial, and other applications. Remote mobility control is realized with the WWW standard SVG/XML. In addition to stereo visual telepresence a WebCam image stream is also displayed in a standard browser window.

Future work will expand the presented MTPTA system with a haptic device in cooperation with the Institute of Automatic Control Engineering, Technische Universität München.

Acknowledgments

The support by the German Research Foundation (DFG) within the Collaborative Research Center SFB453 on "High Fidelity Telepresence and Teleaction" (München) is acknowledged. We would like to thank our colleagues in the telepresence research group at the Technische Universität München, in particular H. Baier, Dr. F. Freyberger, Prof. G. Schmidt. Support by Sedlbauer Inc. for provision of the mobile platform is acknowledged. The SVG/XML and WWW implementation is due to Adrian Zentner and Oliver Schultz.

References

Anderson, R. (1996). Autonomous, Teleoperated, and Shared Control of Robot Systems. In *Proceedings of the IEEE International Conference on Robotics and Automation*, pages 2025–2032, Minneapolis, Minnesota.

Baier, H., Buss, M., Freyberger, F., and Schmidt, G. (2000). Benefits of Combined Active Stereo Visual and Haptic Telepresence. In *Proceedings of the IEEE/RSJ International Conference on Intelligent Robots and Systems IROS*, pages 702–707, Takamatsu, Japan.

Baier, H., Buss, M., Freyberger, F., and Schmidt, G. (2002). Interactive stereo vision telepresence for correct communication of spatial geometry. *The International Journal of the VSP and Robotics Society of Japan*, to appear.

62

Burdea, G. and Zhuang, J. (1991). Dexterous telerobotics with force feedback — an overview. part 1: Human factors, part 2: Control and implementation. *Robotica*, 9:171–178,291–298.

Buss, M. (1994). *Study on Intelligent Cooperative Manipulation.* PhD thesis, University of Tokyo, Tokyo.

Buss, M. and Hashimoto, H. (1993). Skill Acquisition and Transfer System as Approach to the Intelligent Assisting System—IAS. In *Proceedings of the 2nd IEEE Conference on Control Applications*, pages 451–456, Vancouver, British Columbia, Canada.

Buss, M. and Schmidt, G. (1999). Control Problems in Multi-Modal Telepresence Systems. In Frank, P., editor, *Advances in Control: Highlights of the 5th European Control Conference ECC'99 in Karlsruhe, Germany*, pages 65–101. Springer.

Buss, M. and Schmidt, G. (2000). Multi-Modal Telepresence. In *Proceedings of the 17th International Mechatronics Forum, Science and Technology Forum 2000, Plenary*, pages 24–33, Kagawa University, Kagawa, Japan.

Fischer, C., Buss, M., and Schmidt, G. (1996). HuMan-Robot-Interface for Intelligent Service Robot Assistance. In *Proceedings of the IEEE International Workshop on Robot and Human Communication (ROMAN)*, pages 177–182, Tsukuba, Japan.

Hanebeck, U. D. and Saldic, N. (1999). A modular wheel system for mobile robot applications. In *Proceedings of the IEEE RSJ International Conference on Intelligent Robots and Systems (IROS)*, pages 17–23, Kjongju, Korea.

Kaczmarek, K., Tyler, M., and Bach-y-Rita, P. (1994). Electrotactile Haptic Displays on the Fingertips. In *Proceedings of the IEEE International Conference on Engineering in Medicine and Biology*, New York.

Lee, S. (1993). Intelligent sensing and control for advanced teleoperation. *IEEE Control Systems Magazine*, pages 19–28.

Maneewarn, T. and Hannaford, B. (1998). Haptic Feedback of Kinematic Conditioning for Telerobotic Applications. In *Proceedings of the International Workshop on Intelligent Robots and Systems IROS*, pages 1260–65, Victoria, Canada.

Sheridan, T. (1992). *Telerobotics, Automation, and Human Supervisory Control.* MIT Press, Cambridge, Massachusetts.

HAPTIC INTERNET-TELEPRESENCE

Sandra Hirche   Martin Buss

Control Systems Group, Faculty EECS, Technical University Berlin

Einsteinufer 17, D-10587 Berlin, Germany

Hirche@rs.tu-berlin.de, M.Buss@ieee.org

Abstract   This paper discusses Internet-telepresence systems with haptic coupling of operator and environment through a teleoperator. On the basis of known passivation and impedance matching techniques a novel approach to cope with discontinuous time delay variations as they can occur in the Internet is proposed and validated by experimental results.

Keywords: *Telepresence (Teleoperation) System, Haptic Feedback, Internet*

Introduction

Telepresence (teleoperation) systems allow human operators to perform tasks in distant, hazardous, or inaccessible environments as in space, underwater, nuclear plants, etc. The multifaceted variety of applications ranges from tele-medicine, tele-diagnosis for industrial plants over tele-service, tele-manufacturing to tele-shopping. For the human operator in multimodal remote control of e.g. an executing mobile telerobotic system, the feeling of direct interaction with the remote environment is essential to perform fine manipulation tasks. Hence the generation of visual, auditory, and haptic feedback plays a major role in telepresence systems. The feedback technology for the visual and auditory human senses is readily available, whereas haptic systems are much less developed.

The control issues and problems encountered in strongly coupled haptic telepresence systems are the focus of this paper. In the following the basic structure of a telepresence system with haptic feedback and the corresponding communication architecture according to Fig. 1 is discussed. The operator manipulates a force feedback capable (haptic) input device, the so-called Human System Interface (HSI). The position data x_h measured at the HSI is transmitted through a communication channel to the remote teleoperator, which interprets the arriving data as a position reference signal x_t^d (^{d} denotes desired). A local position control loop at the teleoperator side ensures the tracking of

63

G. Hommel and S. Huanye (eds.), The Internet Challenge: Technology and Applications, 63–72.

© *2002 Kluwer Academic Publishers.*

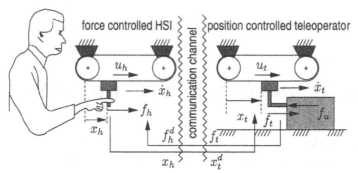

Figure 1.   Fundamental architecture of a telepresence system with haptic feedback

this desired position. On the other hand, the teleoperator force f_t is measured and communicated back to the HSI, where the signal serves as the reference input f_h^d for the local HSI force control loop. The operator experiences the force exerted on the teleoperator by the remote environment, ideally causing the impression to directly interact with the remote environment; this is called *transparency*. The system can be represented by a classical master slave configuration, where the operator together with the HSI act as master and the teleoperator, reacting to the HSI commands, acts as slave. Transparency means that forces, positions, and velocities of HSI and teleoperator are equal, i.e. $x_h = x_t$, $\dot{x}_h = \dot{x}_t$, and $f_h = f_t$.

Stability of telepresence systems with haptic feedback is a major challenge because of time delay and strong coupling between HSI and teleoperator in the global control loop. In contrast to visual and auditory feedback, not only information is transmitted, but the physical energy of the HSI and teleoperator systems are strongly coupled by communication. The most challenging issue is communication delay, experienced by signals in the forward and backward paths. Any control loop with significant time delay causes stability problems. In order to overcome this deficiency a bilateral control law for constant, arbitrary time delay was proposed in [1] transforming the bidirectional communication channel into an ideal lossless transmission line in terms of electrodynamics; see also [2] for an equivalent method using a wave variable formalism and applying the technique of impedance matching, which guarantees not only stable but transparent telepresence even in presence of communication delays larger than 1s. In [3, 4] an overview of some current control algorithms and methods for multimodal telepresence can be found.

The Internet of today as a means for communication provides easy access at low cost. But its unreliable service, variable communication latency, and limited bandwidth is a challenge for application of telepresence systems with high safety requirements. Past work on environments for telerobotics control under variable time delay and package loss proposed to solve the stability problem by either estimating the original data as in [5], by robust control approaches as in [6] or by using reconstruction filters as in [7]. See e.g. [8, 9, 10, 11, 12, 13, 14, 15, 16, 17, 18, 19, 20, 21, 22] for related work.

This paper contributes to the operability of telepresence systems in Internet-like communication networks in the presence of discontinuous time delays and package loss, discussing and experimentally validating a novel approach for numerical optimization of telepresence architectures, see [23] for further details.

For the organization of this paper: In Sec. 1 the prototypical experimental setup of a simple haptic telepresence system is presented. The methods of passivation and impedance matching are discussed in Sec. 2 and 3. Experimental results with discontinuous time delay and package loss are shown in Sec. 4. All experiments were performed with the prototypical telepresence system. Sec. 5 briefly mentions future research directions.

1.   Experimental Setup

The experimental setup consists of two identical single degree of freedom force feedback paddles connected to a PC; the original design of the paddles can be found in [24]. The basic configuration is shown in Fig. 2, where the

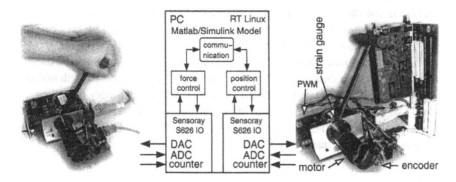

Figure 2.   Basic hardware configuration

teleoperator inserts a PCI expansion card into a motherboard as an example. The paddle DC motor torque is controlled by the PWM amplifier, which operates in current control with the reference given by a voltage from the D/A converter output of the I/O board. The force applied to the paddle lever, attached at the motor axis, is measured through the bending of the lever by a strain gauge bridge, that is attached at the bottom of the lever with the strain being amplified and converted by an A/D converter of the I/O board. The position of the lever, measured by an optic pulse incremental encoder on the motor axis is processed by a quadrature encoder on the I/O board.

The control loops and the model of the communication channel are composed of MATLAB/SIMULINK blocksets; standalone realtime code for RT Linux is automatically generated from the SIMULINK model. All experiments were performed with a sample time $T_A = 0.001$s.

2. Passivation

Common methods for stability analysis from control theory are difficult to apply in telepresence systems for two reasons: (i) the environment, the teleoperator interacts with, is usually unknown; (ii) the human acting as controller in the global closed loop, is very difficult to model. The concept of passivity, applied first in [1], is a useful frame work for analysis and synthesis of such systems. A passive element is one for which, given zero energy storage at $t = 0$, the property

$$\int_0^t P_{in}(t)\,d\tau \geqslant 0 \qquad (1)$$

holds, with $P_{in}(t)$ denoting the power dissipated in the system. If the power dissipation is zero, the system is called *lossless*. If a network is a connection of passive components only, the network itself will be passive and implicitly stable. In Fig. 3 a telepresence system is depicted, see also [23, 25]. A trained person in interaction with the HSI represented by block e) as well as the teleoperator interacting with the remote environment, represented by block a) are assumed to behave passive in [1].

The bidirectional communication channel represented by block b) can be modeled as a time delaying two-port with delay T_1 in the forward path and delay T_2 in the backward path. To verify passivity, scattering theory is applied, as in [1]. It can easily be shown, that the eigenvalues of the scattering matrix are larger than 1, hence the condition for passivity is violated. The communication channel may cause instability.

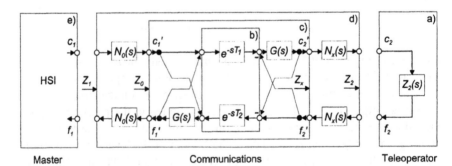

Master   Communications   Teleoperator

Figure 3. Block diagram of passivated, impedance matched telepresence system

The passivation method mentioned above modifies the communication block by introducing new feedback paths as illustrated by block c) in Fig. 3 with the filter $G(s) = 1$. The performance of the passivated system is evaluated by comparing the impedance, felt at the operator side with the teleoperator impedance; both equal in case of transparency.

Impedance is defined by the ratio of effort (force f) and flow (velocity v) as

$$Z = \frac{f}{v}.$$

In the more generalized definition of mechanical impedance, which is applied here, the velocity is replaced by the position.

For transparency the transmitted impedance Z_0 should be equal (or as close as possible) to the teleoperator impedance Z_2, resulting in a realistic haptic impression. The Bode plot in Fig. 4 shows the frequency dependent force response of the teleoperator to a position command from the HSI in case of a constant time delay. The dashed line represents the teleoperator impedance Z_2, which is the nominal or desired impedance. After passivation the transmitted impedance Z_0 (dash-dotted line) strongly deviates from the nominal impedance, hence transparency is not achieved. The experimentally obtained position versus time plots of the HSI and the teleoperator in Fig. 5(a) validate that; unstable behavior causes a safety shutdown.

3.   Impedance matching

If the terminating impedance Z_x of the modified transmission line differs from the teleoperator impedance Z_2, then the wave energy received from the transmission line is not absorbed by the teleoperator system, causing unstable behavior. An impedance matching technique, first introduced in [26] aims to adapt the impedances of the teleoperator and the transmission line.

According to [25] the passive transmission line needs to be extended by dynamic filters N_x and N_0

$$N_x(\omega) = \frac{1}{\sqrt{Z_2(\omega)}}, \quad N_0 = \frac{1}{N_x} \tag{2}$$

as illustrated in Fig. 3d). In theory ideal transparency $Z_1 = Z_2$ as defined in [27] for arbitrary communication delays is reached. This not only requires the exact model of the teleoperator, but also the knowledge of the environmental impedance. In practice the ideal filters are not available as it requires the computation of the square root of an arbitrary transfer function according to (2), which can only be approximated. A novel approach to obtain this approximation through numerical optimization is discussed in the following, see also [23].

68

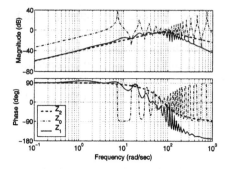
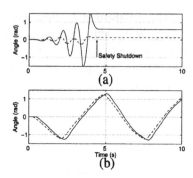

Figure 4. Bode plots teleoperator impedance Z_2, transmitted impedances Z_0 after passivation and Z_1 after impedance matching

Figure 5. HSI (dashed) and teleoperator (solid) position for $T_1 = T_2 = 0.2s$ after passivation (a) and and after impedance matching

The quality of impedance matching can be expressed in terms of a distance between the current terminating impedance $Z_x(N_x, \omega)$ and the ideal terminating impedance $Z_{x,id} = 1$ in the Nyquist-Diagram. The aim of optimization is to minimize this distance. For the given free space motion scenario with constant time delay a leadlag filter of second order achieves good results. In order to cancel high frequency disturbances, a low-pass filter $G(s)$ with a cut frequency of 50Hz is inserted, as illustrated in Fig. 3d). The optimized impedance matched system provides superior performance in simulation as well as in experimental evaluation. The transmitted impedance $Z_1(\omega)$, presented in Fig. 4 only slightly deviates from the nominal impedance $Z_2(\omega)$ over a large bandwidth. This result is verified by the tracking results in Fig. 5(b).

4. Discontinuous Time Delay and Package Loss

The approaches mentioned above aim to stabilize and improve performance for a communication channel with constant time delay. In packet-switched, digital communication networks of today, as e.g. the Internet, variable time delay and also package loss need to be considered as additional performance degrading or even destabilizing influence. From the applications point of view the data transmission quality is sufficiently described by the parameters time delay, jitter, package loss, and bandwidth.

Haptic data, sent in packets from the HSI to the teleoperator and back experience a variable, network load dependent, discontinuous time delay. The variance, caused by random queuing delays in network devices, is called jitter. As soon as the capacity in the queue of a network device is exceeded, package loss occus. Jitter as well as package loss cause the distortion of the haptic sig-

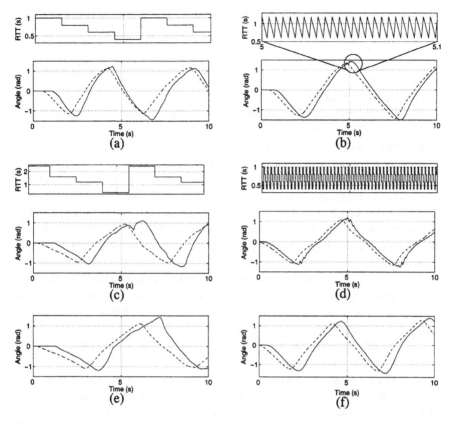

Figure 6. Impact of discontinuous time delay on position tracking of HSI and teleoperator

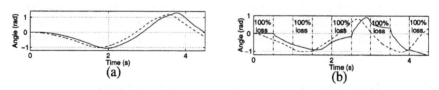

Figure 7. Impact of package loss distribution on position tracking of HSI and teleoperator

nal, introducing high frequency noise if always youngest data is passed to the HSI/teleoperator subsystems.

The effect of the above mentioned bandlimiting filter $G(s)$ in combination with the optimized impedance matching filters N_x and N_0 is examined in several scenarios, contributing to some typical situations in Internet-like communication networks. In the first scenario, see Fig. 6(a), the roundtrip time is constant for 1.5s before steps occur. Naturally, the lag between HSI and teleoperator position rises, hence performance deteriorates with higher roundtrip delay. The system remains stable, even if the time delay increases in steps, see Fig. 6(b). The now higher distortion of the signal compared to the first scenario is decreased, if the cut frequency of the filter $G(s)$ is lowered from 50Hz to 10Hz, as illustrated in Fig. 6(c). Additional phase lag between HSI and teleoperator position is thereby introduced.

In order to lower protocol overhead and bandwidth requirements packets of 100 samples are collected and then sent; therefore the delay is constant for 100 samples (i.e. 100ms). The teleoperator position signal is slightly noisy, especially when delay steps occur at a time where the position change is large, see Fig. 6(d). The impact can be decreased by lowering the cut frequency of the filter $G(s)$ to 10Hz, as illustrated in Fig. 6(e). Even if time delay changes with every single sample data, the system performs well, see Fig. 6(f). The hereby introduced high frequency noise, but little magnitude is completely filtered out by $G(s)$ with a cut frequency of 50Hz, so that no major impact on performance is noticeable.

The impact of jitter on the performance mainly depends on the resulting frequency and magnitude of noise, which in turn depends on the value of jitter and the shape of the communicated haptic signal. Low-pass filtering smoothens the jitter distorted signals, but induces additional time lag. Instability did not occur in any of the performed experiments, however a formal proof of stability margins remains future work.

The impact of single package loss on the telepresence system mainly depends on the amount of information transmitted in one packet. Usually a single data loss neither changes the subjective haptic impression nor does it destabilize the system, as visualized in Fig. 7(a), where every second data (50% of data) was lost. Far more crucial is the loss of a number of consecutive data in a stream. In Fig. 7(b) the results are shown for 50% package loss, with bursts of 500 consecutive samples being lost. Hence not only the amount but the distribution of package loss is important for the performance of the haptic teleoperation system.

5. Improvements by Quality-of-Service

Low cost and easy access are the main reasons to choose the Internet as a communication infrastructure in telepresence, even if the "best effort" service model as applied today does not adequately meet the realtime requirements for stable telepresence with haptic feedback. With the transitioning from IPv4 to IPv6, there is a chance of implementing protocols, that guarantee certain performance characteristics and therefore provide the so-called *Quality-of-Service*-Model. Considering telepresence systems with the required performance mainly depending on the task and the environment, the data transmission quality, expressed in terms of communication delay, jitter, package loss, and bandwidth, may act as new tuning parameters for telepresence system performance.

6. Conclusions

In this paper control of haptic teleoperation systems operated in the Internet has been discussed. Based on passivity arguments and optimized impedance matching a novel approach has been proposed. Experimental results with a prototypical haptic teleoperation system have confirmed the validity of the approach, which also looks promising in Internet-like communication networks with discontinuously changing time delay and package loss. Future work is to generalize this approach to Quality-of-Service communication infrastructures.

Acknowledgment

This work has been partially supported by the "IBB-Zukunftsfond Berlin, Projekt Neue Generation leittechnischer Systeme (mobile und virtuelle Leittechnik)".

References

[1] R. Anderson and M. Spong, "Bilateral Control of Teleoperators with Time Delay," *IEEE Transaction on Automatic Control*, vol. 34, pp. 494–501, 1989.

[2] G. Niemeyer and J.-J. Slotine, "Stable Adaptive Teleoperation," *IEEE Journal of Oceanic Engineering*, vol. 16, pp. 152–162, January 1991.

[3] M. Buss and G. Schmidt, "Control Problems in Multi-Modal Telepresence Systems," in *Advances in Control: Highlights of the 5th European Control Conference ECC'99 in Karlsruhe, Germany* (P. Frank, ed.), pp. 65–101, Springer, 1999.

[4] S. Salcudean, "Control for Teleoperation and Haptic Interfaces," in *Lecture Notes in Control and Information Sciences 230: Control Problems in Robotics and Automation* (B. Siciliano and K. Valavanis, eds.), pp. 51–66, London: Springer, 1998.

[5] A. R. R. Luck and Y. Halevi, "Observability under Recurrent Loss of Data," *AIAA Journal of Guidance, Control and Dynamics*, vol. 15, pp. 284–287, 1992.

[6] Y.-P. Huang and K. Zhou, "Robust Stability of Uncertain Time-Delay Systems," *IEEE Transactions on Automatic Control*, vol. 45, pp. 2169–2173, 2000.

[7] G. Niemeyer and J. E. Slotine, "Towards Force-Reflecting Teleoperation Over the Internet," in *Proceedings of the IEEE International Conference on Robotics and Automation*, (Leuven, Belgium), pp. 1909–1915, 1998.

[8] H. Baier, M. Buss, and G. Schmidt, "Control Mode Switching for Teledrilling Based on a Hybrid System Model," in *Proceedings of the IEEE/ASME International Conference on Advanced Intelligent Mechatronics AIM'97*, (Tokyo, Japan, Paper No. 50), 1997.

[9] W. Book, H. Lane, L. Love, D. Magee, and K. Obergfell, "A Novel Teleoperated Long-Reach Manipulator Testbed and its Remote Capabilities via the Internet," in *Proceedings of the IEEE International Conference on Robotics and Automation*, (Minneapolis, Minnesota), pp. 1036–1041, 1996.

[10] M. Buss and G. Schmidt, "Multi-Modal Telepresence," in *Proceedings of the 17th International Mechatronics Forum, Science and Technology Forum 2000*, (Kagawa University, Kagawa, Japan), pp. 24–33, 2000.

[11] H. Lee and M. Chung, "Adaptive Control of a Master-Slave System for Transparent Teleoperation," *Journal of Robotic Systems*, vol. 15, pp. 465–475, 8 1998.

[12] W. Ferrell and T. Sheridan, "Supervisory control of remote manipulation," *IEEE Spectrum*, pp. 81–88, October 1967.

[13] W. Ferrell, "Delayed force feedback," *IEEE Transactions on Human Factors*, vol. HFE-8, pp. 24–32, September 1966.

[14] B. Hannaford, "A design framework for teleoperators with kinesthetic feedback," *IEEE Transactions on Robotics and Automation*, vol. 5, pp. 426–434, August 1989.

[15] K. Hirai and Y. Satoh, "Stability of a System with Variable Time Delay," *IEEE Transactions on Automatic Control*, pp. 552–554, 1980.

[16] K. Kosuge, T. Itoh, and T. Fukuda, "Scaled Telemanipulation with Communication Time Delay," in *Proceedings of the IEEE International Conference on Robotics and Automation*, (Minneapolis, Minnesota), pp. 2019–2024, 1996.

[17] K. Kosuge, H. Murayama, and K. Takeo, "Bilateral Feedback Control of Telemanipulators via Computer Network," in *Proceedings of the IEEE/RSJ International Conference on Intelligent Robots and Systems IROS*, (Osaka, Japan), pp. 1380–1385, 1996.

[18] G. Leung, B. Francis, and A. Apkarian, "Bilateral controller for teleoperators with time delay via μ-synthesis," *IEEE Transactions on Robotics and Automation*, vol. 10, pp. 105–116, 1993.

[19] P. Arcara and C. Melchiorri, "Control Schemes for Teleoperation with Time Delay: A Comparative Study," *Robotics and Autonomous Systems 950*, pp. 1–16, 2001.

[20] T. Sheridan, *Telerobotics, Automation, and Human Supervisory Control*. Cambridge, Massachusetts: MIT Press, 1992.

[21] T. Sheridan, "Telerobotics," *Automatica*, vol. 25, no. 4, pp. 487–507, 1989.

[22] M.-Y. Shi, "Designing Force Reflecting Teleoperators with Large Time Delays to Appear as Virtual Tool," in *Proceedings of the IEEE International Conference on Robotics and Automation*, pp. 2212–2218, 1997.

[23] S. Hirche, "Control of Teleoperation Systems in QoS Communication Networks." Diploma Thesis, Control Systems Group, Technical University Berlin, 2002.

[24] H. Baier, M. Buss, F. Freyberger, J. Hoogen, P. Kammermeier, and G. Schmidt, "Distributed PC-Based Haptic, Visual and Acoustic Telepresence System—Experiments in Virtual and Remote Environments," in *Proceedings of the IEEE Virtual Reality Conference VR'99*, (Houston, TX), pp. 118–125, 1999.

[25] H. Baier, M. Buss, and G. Schmidt, "Stabilität und Modusumschaltung von Regelkreisen in Teleaktionssystemen," *at—Automatisierungstechnik*, vol. 48, pp. 51–59, Februar 2000.

[26] G. Niemeyer and J. E. Slotine, "Using Wave Variables for System Analysis and Robot Control," in *Proceedings of the IEEE International Conference on Robotics and Automation*, pp. 1619–1625, 1997.

[27] D. Lawrence, "Stability and Transparency in Bilateral Teleoperation," *IEEE Transactions on Robotics and Automation*, vol. 9, pp. 624–637, October 1993.

A SIMULATION FRAMEWORK FOR SUPPLY CHAIN MANAGEMENT IN AN E-BUSINESS ENVIRONMENT

Alexander Huck, Michael Knoke, Günter Hommel

Technische Universität Berlin, Germany

Real-Time Systems and Robotics

{ahuck, knoke, hommel}@cs.tu-berlin.de

Abstract   The information age has driven a paradigm shift from traditional mass marketing of finished goods to online marketing of built-to-order products. With the customer in control, companies are forced to rethink their management of supply chains permanently. Modern supply chain management performance is guided by variability – or the ability to manage change. Supply chain simulation helps companies to make better forecasts and adapt to the challenges of the future. This paper presents a Petri-net based framework for supply chain simulation which is embedded into a distributed environment. It describes the problem domain and new software technologies, assesses their impact on simulation integration and interoperation, and provides a representative scenario of a complex vehicle manufacturing system.

Keywords: supply chain modelling, petri net simulation, e-business environment

Introduction

The term *supply chain* encompasses the web of interconnected relationships between sales channel, distribution, warehousing, manufacturing, transportation, and suppliers. This network covers all activities from raw materials to the final consumer. Without any coordination efforts, each organization in the chain has its own agenda and operates independently from the others, which can lead to an unstable, oscillating behavior of the network as each organization in the supply chain seeks to solve the problem from its local perspective. This phenomenon is known as the *bullwhip effect* and has been observed across most industries. *Sup-*

G. Hommel and S. Huanye (eds.), The Internet Challenge: Technology and Applications, 73–82.

© 2002 *Kluwer Academic Publishers.*

ply chain management tries to overcome these problems by utilizing a holistic view and emphasizing the need for collaboration to optimize the whole system. As such, supply chain management is the process of designing, planning, and implementing change in the structure and performance of the total material flow in order to generate increased value, lower costs, enhance customer service, and yield a competitive advantage [Rich01].

In the automotive industry, all activities that occur between a consumer's decision to purchase a vehicle and the final delivery of that vehicle to the customer are subsumed under the term *Order-to-Delivery* (OTD). Automotive companies try to reduce their OTD-time to a few days while selling vehicles that are tailored to the customer's wishes. This has to be done without building large costly pools of inventory, and by informing all tiers of the supply chain instantly and simultaneously of customer demand changes.

In this situation, supply chain simulation can help to make informed decisions about business strategies. It permits the evaluation of operating performance prior to the implementation of a system as well as the comparison of various operational alternatives without interrupting the real system. Based on accumulated operation data, simulation can help to synchronize supply plans, reduce the occurrence of exceptions, diagnose problems, and evaluate possible solutions. This can turn supply chain complexity into an important competitive advantage.

Petri nets are supporting modelling and simulation of complex systems from the scratch [ZFH01]. They provide a graphical representation and are able to represent discrete events as well as stochastic timing [BP95]. Our simulation framework uses a new class of *colored* Petri nets [Jen92] that has been enhanced with some additional features such as the capability to access external data repositories. The implementation is based on the tool TimeNET [ZFGH00, GKZH92] which has good modelling and visualizing support.

This paper is structured as follows: Section 1 introduces a fictitious vehicle manufacturer and its extensive supply chain as a sample application environment for our simulation framework. Section 2 gives a description of our simulation framework including the network connectivity of all components. Section 3 introduces a new class of colored Petri nets that is used as the modelling language of the simulation framework. Finally, Section 4 presents a conclusion and shows directions for further studies.

1. Application Environment

More and more automobile corporations utilize *logistics control centers* in their supply chain to realize Order-to-Delivery (OTD) and achieve the International Environmental Management System (EMS) standard ISO 14001 [Cas99]. This standard is internationally recognized like the Quality Management System (QMS) standard described in ISO 9001 and certifies that companies are implementing a strong and effective environmental management program. It specifies requirements for establishment, implementation and assessment of an environmental policy.

The logistics control center coordinates the management of all service providers and supervises the interaction between suppliers, plants, dealers, customers, and all other parts of the supply chain. It is responsible for inbound material, finished vehicles, and aftermarket products. Primary objective is to eliminate silos creating common, integrated solutions across logistics networks.

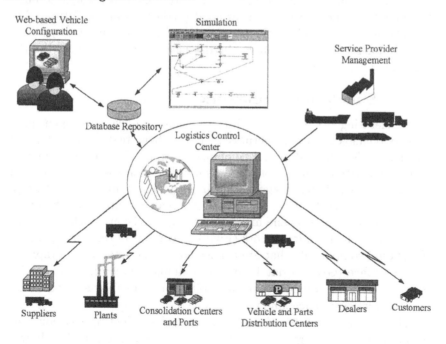

Figure 1. Supply Chain Visibility and Management

The automotive corporation in our sample application environment (Figure 1) provides a web site that allows customers to configure and purchase their car online. [1] They are able to visit different vehicle design and option proposals and can change them to match their favored configuration. The customer behavior is recorded to a database and yields valuable information for simulation runs.

When a customer has made a purchase decision, the logistics control center schedules the corresponding vehicle for assembly. Parts and component modules are delivered sequentially, as needed, from the suppliers to the plants and consolidation centers. Vehicle and parts distribution centers are responsible for spot delivery to dealers and for exchange of vehicles between dealers. The flow of inventory at every point in the logistics process is monitored by the logistics control center and dumped to a database.

It is the job of the simulation to test and evaluate different strategies for the work of the logistics control center. To fulfil this task, the simulator has access to a database repository that contains information about the customer behavior and the flow of inventory. By combining knowledge from the past with future expectations, it is possible to build a simulation model that produces good predictions of the future. The results are used to adapt the supply chain management strategy used by the logistics control center. In the next chapter, we describe the simulation framework in detail.

2. Simulation framework

Accelerating OTD requires virtual business models with open, integrated technology infrastructures to make them real. Internet-based simulation is a first step and combines the powerful scenario building and visualization features of a good simulator with the portability and visibility of a web-based application [Lin00]. The computer model represents a "virtual factory" for automotive manufacturing which can be easily used by engineers and managers to make decisions quickly and reliably. Our simulation framework (Figure 2) consists of three independent functional components that are connected via the internet: a graphical user interface (GUI), a database, and a simulation engine. This architecture offers the advantages of data collection from a variety of sources, platform independence, and enterprise-wide visibility.

The GUI is a web-based and platform-independent application that enables users to design, test, and evaluate a simulation model. It was derived from the Petri net tool TimeNET and enhanced with new components. The model structure is designed in a graphical editor that allows quick changes by simply using the mouse. In addition the GUI provides an interface to assign database entries to model entities and automatically generates SQL (Structured Query Language) queries that

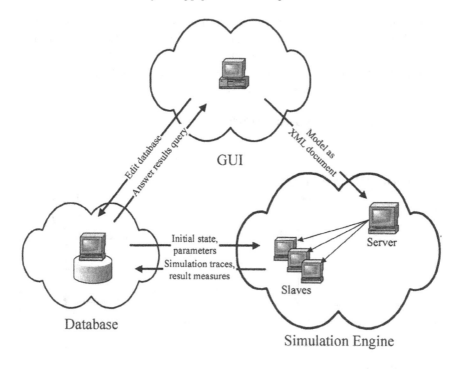

Edit database

Answer results query

Model as XML document

GUI

Initial state, parameters

Simulation traces, result measures

Server

Slaves

Database

Simulation Engine

Figure 2. Overview of the simulation framework

are embedded in the model. In doing so, it is easy to use data from the company's databases in the simulation. The resulting model is stored in an XML document.

A simulation run is started by sending the XML model document containing all structural information to the simulation server. The server then generates C++ code from the model document, which is compiled and linked together with a simulation kernel library. The resulting program is executed on a number of hosts (slaves). At the beginning of a simulation run, the initial model state as well as model parameters are read from a database. In order to debug a newly built model, the simulator is capable to run in interactive mode with support of the GUI, in which the user has complete control over all decisions. If desired by the user, the slaves can produce output in form of simulation traces, which are recorded to a database for offline analysis. This could be, for example, a list of all cars with their production dates. Alternatively, the simulation engine is able to perform an online analysis of numeric performance measures, e.g. the mean number of cars produced per day. These measures are collected from the slaves and accumulated by the master. Again, the results are stored in a database. The simulation progress and the final results can be simultaneously displayed by the GUI running on different computers. It provides a visualization by plotting the change of the performance measures over time.

3. Modelling Language

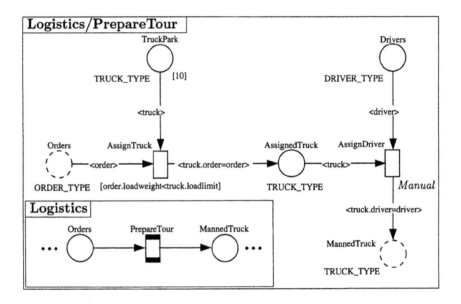

Figure 3. A Sample Model

As the modelling language, a new class of high-level Petri nets is used. High-level Petri nets differ from classical Petri nets by some well-known extensions [Aal94]: They use typed (colored) tokens to model data, they possess a timing concept to describe the temporal behavior of a system, and they provide a notion of hierarchy. Our newly devised class of Petri nets implements all of these concepts plus some extra features that further facilitate the modelling of complex systems. This section introduces the net class by means of the small example shown in Figure 3, which can be thought of as part of a logistics network. It shows how shipment orders and truck drivers are assigned to trucks before the trucks are loaded and start their tour.

High-level nets represent objects and their attributes (like the trucks in the example above) by *typed tokens*. In our net class, a token type may consist of

- simple data types like integer or string,

- other types of tokens and references to tokens,

- lists of the above.

The token types are mapped onto database tables so that the marking of a net can be stored and restored. In order to model an "is-a"-relationship between types, it is possible to derive token types by extension from other token types. In the example, a truck would be described by a reference to the shipping order that the truck has been assigned to, a reference to truck driver, the maximum load limit, and a list that describes the load: TRUCK_TYPE = { ORDER_TYPE* order, DRIVER_TYPE* driver,
            INTEGER loadlimit, LOAD_TYPE[] load } .

Places model buffers and other token containers. A place has a fixed token type assigned to it and can hold tokens of that type only, e.g. the TruckPark may only contain tokens of type TRUCK_TYPE. It is possible to assign a capacity to a place which can also be read from a database. Each place has a fixed token access strategy to speedup the simulation process, that defines the way that tokens are stored inside the place. With a FIFO (first in, first out) strategy, only the oldest token of a place is available for transition firing and all other tokens are ignored. Other strategies are LIFO (last in, first out) and random access.

Substitution Transitions are used to create a hierarchy. They are placeholders for a refining submodel. In the example, the substitution transition PrepareTour on the Logistics page is refined by a submodel that describes the details of assigning orders and drivers to trucks. Each substitution transition has an additional integer attribute "replication" which defines the number of times the associated submodel will be instantiated. Inside the submodel, the current number is known under the predefined variable "id", e.g. to have place capacities based on this number.

Transitions model the activities of the system. Each transition has a firing delay that specifies the time that elapses before an enabled transition fires. It is given by a firing time distribution. The process of firing is divided in two parts. Based on the input arc inscriptions a valid combination of tokens is removed from the input places and is put to the output places on the basis of output arc inscriptions. The selection of valid combinations of input tokens (bindings) is performed by boolean expressions called guards. The transition AssignTruck in Figure 3 has a boolean guard that accepts only pairs of tokens (order, truck) where the weight of the load as stated in the shipping order (order.loadweight) doesn't exceed the load limit of the truck (truck.loadlimit). In addition to boolean guards, special time guards are available to define the enabling of a transition depending on the current simulation time.

A transition may be replaced completely or in parts by a manual definition, namely a C++ code fragment that follows a standardized form. This is intended for cases where the normal graphical modelling capacities of the Petri net class aren't sufficient. In the example, the AssignDriver transition is a manual transition, because the assignment of drivers to trucks follows a complex scheduling strategy that would be hard to code into a Petri net.

The flow of tokens through the net is described by *arc inscriptions*. Input arcs (arcs from a place to a transition) have a inscription that describes which tokens are necessary to fire a transition, and which tokens should be removed from the input places when the transition fires. They carry a variable name in pointed brackets whose type is defined by the corresponding place. The variable names are used by the guard expression of the transition and by the output arcs. These arcs (from a transition to a place) describe which tokens are to be added to the output places when the corresponding transition fires. It is possible to move, copy, modify and create new tokens. In the sample model (Figure 3), the transition AssignTruck has two input arcs that are labelled with the variable names order and truck respectively. To check if the transition is enabled, the set of all variable assignments that satisfy the transition guard is generated, i.e. the set of orders and trucks that match the load weight constraint. When the transition fires, one of the valid bindings is randomly selected. Then, the corresponding truck token is removed from the TruckPark place, and the order token is removed from the Orders place. The inscription on the output arc describes how to move the truck token to the AssignedTruck place and modify it by assigning the order token to the order attribute of the truck token.

Based on the concept of colored Petri nets, our newly devised net class has been tailored to fit the needs of supply chain simulation. With the option to integrate C++ code into the model we elevate the modelling power to handle situations where Petri nets alone are not sufficient. This feature also allows for integration of external programs. With database access directly from net components, it is possible to dynamically link to different external data repositories.

4.   Conclusion and Outlook

This paper describes a new simulation framework for the simulation of supply chains. As a sample application framework, we present a vehicle manufacturer facing the challenge to sell highly customized vehicles while minimizing the OTD-time. Management of an extensive supply chain in the automotive industry needs simulation for the evaluation of different strategies. This can optimally be achieved using real data. Our simulation system provides database integration to be able to access data collected from different sources in the company. Its distributed architecture makes it also possible to inspect simulation results from different locations using the web. As the modelling language, we use a class of colored Petri nets with specific properties that facilitate the task of supply chain simulation.

Future work will include distributed simulation on cluster systems and may include investigating security aspects of the network connections. Both are important steps to introducing forecast simulation in manufacturing systems with high Business-to-Business and Business-to-Customer relationships, where fast simulation run time and easy modelling of complex supply processes are required.

Notes

1. Some real life examples for online configuration and purchasing facilities are

■ "BuyPower" (General Motors, http://www.gmbuypower.com),

■ "Car.Configurator" (Mercedes Benz, www.car-configurator.mercedes-benz.com) and

■ "Build your Volvo" (Volvo, http://new.volvocars.com/build).

References

[Aal94] W.M.P. van der Aalst, "Putting Petri nets to work in industry," Computers in Industry, 25(1):45-54, 1994.

[BP95] R. Bastide, and P. A. Palanque, "A Petri Net based Environment for the Design of Event-driven Interfaces," Application and Theory of Petri Nets (1995), 66-83.

[Cas99] J. Cascio, "The Iso 14000 Handbook," American Society for Quality, ISBN: 0873894405, 1999.

[GKZH92] R. German, C. Kelling, A. Zimmermann, and G. Hommel, "TimeNET - A Toolkit for Evaluating Non-Markovian Stochastic Petri Nets," Performance Evaluation 24 (1995) 69-87.

[Jen92] K. Jensen, "Coloured Petri Nets: Basic Concepts, Analysis Methods and Practical Use," EATCS Monographs on Theoretical Computer Science (Springer Verlag, 1992).

[Lin00] Bo Lindstrom, "Web-based interfaces for simulation of coloured Petri net models," International Journal on Software Tools for Technology Transfer, Volume 3, Issue 4, 405-416.

[Rich01] N. Rich, "An Executive Introduction To Supply Chain Management," Deloitte & Touche Senior Executive Series of Briefing Papers, http://www.deloitte.co.uk/sectors/manufacturing/publications.html.

[ZFGH00] A. Zimmermann, J. Freiheit, R. German, and G. Hommel, "Petri Net Modelling and Performability Evaluation with TimeNET 3.0," 11th Int. Conf. on Modelling Techniques and Tools for Computer Performance Evaluation (Schaumberg, Illinois, USA, 2000) 188-202, LNCS 1786.

[ZFH01] A. Zimmermann, J. Freiheit, and A. Huck, "A Petri net based design engine for manufacturing systems," Int. Journal of Production Research, special issue on Modeling, Specification and Analysis of Manufacturing Systems 39 (2) (2001) 225-253.

PERFORMANCE COMPARISON OF 802.11B WLAN TRANSMISSION MODES

Armin Heindl, Günter Hommel
TU Berlin, Fakultät Elektrotechnik und Informatik
Institut für Technische Informatik und Mikroelektronik
Einsteinufer 17, 10587 Berlin
heindl,hommel@cs.tu-berlin.de

Abstract   In 1999, IEEE supplemented the 802.11 standard for wireless local area networks (WLANs) in order to provide for higher payload data rates. This higher-speed physical layer (PHY) extension in the 2.4 GHz band, approved as document IEEE 802.11b, specifies two advanced modulation schemes (namely CCK and PBCC) for the direct sequence spread spectrum (DSSS) technology, thus enabling data rates of 5.5 and 11 Mbps in addition to 1 and 2 Mbps. Together with the optional Short Preamble mode (instead of the conventional Long Preamble), an 802.11b WLAN may transmit in seven different modes[1]. Based on the fundamental contention-based access mechanism – the distributed coordination function (DCF), which allows two different handshaking techniques –, we compare the performance of the resulting transmission modes. To this end, we adapt a previously published stochastic Petri net model and evaluate it by simulation.

Keywords: IEEE 802.11b wireless LANs, stochastic Petri nets, simulations

Introduction

As an extension of the third generation (3G) cellular networks (i.e., UMTS and CDMA2000) into devices without direct cellular access, wireless LANs play an important role in Internet development. WLANs operating at 2.4 GHz with 11 Mbps in accordance with the IEEE 802.11b standard [4] can already be found on the market for localized data delivery at affordable prices. Still, in an interworking of different technologies in heterogeneous networks, thorough performance analyses are required, for example, to choose the best connectivity among alternative solutions

---

[1]Note that short preambles preclude the 1 Mbps data rate.

G. Hommel and S. Huanye (eds.), The Internet Challenge: Technology and Applications, 83-92.
©2002 *Kluwer Academic Publishers.*

for a user's purpose. In this context, we investigate the performance options envisioned for 802.11b WLANs.

This paper builds on previously published stochastic Petri nets (SPNs) for 802.11 WLANs [6], which have proven to be a flexible modeling formalism. The 802.11b extension may conveniently be covered by modifying PHY parameters of these models. In the literature, many simulation studies on the performance of 802.11(b) protocols have been presented [1, 2, 8, 7]. Most often, specific implementation details are not outlined in the publications. In contrast, SPNs allow to capture relevant performance aspects in a concise and graphical form. We use the stochastic Petri net language (SPNL, [5]) with its ability to structure complex models into modules. Non-exponential timing enhances modeling power, while at the same time the SPNL models can be efficiently evaluated by the software tool TimeNET [9].

In Sec. 1, the 802.11(b) standards are briefly reviewed with emphasis on the features related to the b-supplement. After a description of the SPN model in Sec. 2, simulation results (in terms of throughput and mean waiting time) based on this model are presented in Sec. 3.

1.   802.11 DCF and PHY extensions of 802.11b

The 802.11 standard specifies in detail the medium access control (MAC) and the physical layer for WLANs operating in the 2.4-2.5 GHz band. In this paper, we consider so-called *Independent Basic Service Sets* (IBSSs) as the basic type of an IEEE 802.11 WLAN. In such an *ad-hoc network*, a finite number of stations – typically only a few – communicate directly in a peer-to-peer manner within the coverage area with DCF being the fundamental MAC access scheme. The DCF mechanism is a variant of carrier sense multiple access with collision avoidance (CSMA/CA). Opposed to wired LANs, collisions – as they occur, if two or more stations initiate a handshake at almost the same time (*vulnerable period*) – cannot be detected immediately. Instead, the lack of a positive acknowledgment from the receiver of a frame indicates a failed transmission to the source station. As in [6], we assume perfect channel sensing and ideal channel conditions. Particularly, each station hears any other station and all stations of a cell contend for the same channel. Using the standardized MAC layer, 802.11b merely affects the PHY settings to improve performance. The following two subsections are dedicated to the DCF and these PHY alterations, respectively.

1.1   The Distributed Coordination Function

In this summary, we ignore some mandatory features of the standard (like EIFS and TSF), because they have been shown to only have a marginal impact on system performance in the considered scenario (see [6]). Two versions of the DCF are defined: *Basic Access* (BA) based on two-way handshaking and *Request-To-Send/Clear-To-Send* (RTS/CTS) based on four-way handshaking. A station can use RTS/CTS for data frames exceeding a configurable threshold. In both cases only the first packet has to contend for the medium and the access is based on two time periods: the *DCF Interframe Space* (DIFS) and *Short Interframe Space* (SIFS) with SIFS < DIFS.

Basic Access functions as follows:

1 A station wishing to transmit a directed data frame senses the channel.

2 If the channel has been idle for longer than a DIFS, it transmits the frame and waits for a positive acknowledgment (ACK).

3 The station goes into backoff if

(a) the channel has not been idle for a period of a DIFS.

(b) no ACK has arrived in time (corresponding to a collision).

(c) the frame is consecutive to a previous transmission of the same station (to prevent channel seizure).

After successful reception of the data frame, the receiving station waits for a period equal to SIFS and sends the ACK.

In RTS/CTS, two more packets are exchanged in advance: a short RTS packet by the sender, which must be acknowledged with a CTS packet by the intended receiver, informs of an upcoming data frame. The gaps between successive transmissions are equal to SIFS. Since only the short RTS contends for the medium, collisions (indicated by the lack of the CTS response) waste less bandwidth, not to speak of the effectiveness of RTS/CTS in the presence of hidden terminals.

In both handshakes, there are two ways for a station to discover that its transmission has failed so that the backoff procedure is invoked: either it does not receive the expected response frame (ACK or CTS, respectively) within a specified timeout, or it senses the transmission of a different packet on the channel after its own transmission.

A slotted binary exponential backoff scheme is applied. As soon as the channel is monitored idle for a DIFS, the station selects a backoff time composed of a random number (*backoff counter*) of slot times. With no medium activity indicated for the duration of such a slot of size *aSlot-Time*, the backoff counter is decremented. It is temporarily suspended or "frozen" for periods when a transmission is detected on the channel. Before the backoff procedure is resumed, the channel must have been sensed idle for the duration of a DIFS period. Whenever the backoff counter reaches zero, the transmission commences. The backoff time is

$$BackoffTime = BackoffCounter \times aSlotTime \,,$$

where *BackoffCounter* is a uniformly distributed integer in $[0, CW]$. The contention window CW is an integer equal to $(aCWmin + 1) \cdot 2^{bc} - 1$, where $aCWmin$ is the initial value and bc is a variable which is initialized with zero and incremented before a repeated backoff procedure for a pending frame. The quantity bc can grow up to a maximum value $bcmax$ corresponding to $aCWmax$; afterwards it remains unchanged until it is reset to zero by the next successful transmission.

The *vulnerable period*, during which an ongoing transmission may be corrupted by other stations, because their MAC layers can only recognize that transmission with some delay, is a critical parameter (see Fig. 1 for its composition). Generally, *aSlotTime* should be at least as large as the vulnerable period.

1.2 PHY extensions

Although 802.11 defines three physical layers, the b-supplement concentrates on the spread spectrum technology DSSS. Besides various optional capabilities, 802.11b introduces three performance-relevant features: Most important, new modulation schemes provide payload data rates of 5.5 and 11 Mbps (in addition to the still possible bit rates 1 and 2 Mbps). Second, to further increase system throughputs, the transmission time of the *PHYHeader* – composed of the *PLCPPreamble* and *PLCPHeader* – may be halved for bit rates $B = 2, 5.5, 11$. This is achieved by reducing the *PLCPPreamble* to half its size and transmitting the *PLCPHeader* at twice the minimum bit rate of $B = 1$ usually used for *PHYHeader*. Third, the maximum payload of a data packet *MaxFrameBody* is limited to 4061 bytes (instead of 8157 bytes).

The so-called Long Preamble mode ensures that the physical layers of both 802.11 and 802.11b can coexist in the same WLAN. Combining the different data rates with the alternative Long and Short (not for $B = 1$) Preamble modes and with either BA or RTS/CTS distinguishes 14 transmission modes, which will be assessed in Sec. 3.

Fig. 1 summarizes all parameters employed in the SPN model of the next section – grouped according to their PHY-(in)dependence. The

```
module DSSSpar;

(* all times in us *)

(* parameters for different transmission modes *)
parameter B = 11;              (* BitRate in Mbps; possible values: 1,2,5.5,11 *)
          ShortFactor = 1;     (* 1=Long Preamble mode, 2=Short Preamble mode (precludes B=1) *)
          Handshake = 1;       (* 1=Basic Access, 2=RTS/CTS *)

(* PHY-independent parameters *)
parameter V=0.1;               (* virtual load, varied from 0.1 to 10 for each scenario *)
          N=10;                (* number of stations in cell *)
          K=2;                 (* max # of packets in MAC layer of single station *)

          MaxFrameBody=4061*8; (* in bits, maximum data payload *)
          L = MaxFrameBody/2;  (* in bits, mean data payload assuming uniform distr. *)
          lambda = V * B / (N * L); (* packet arrival rate at a single station *)

          MinBitRate = 1;      (* in Mbps; minimal bit rate, e.g., for PHYHeader transmission *)
          aAirPropagationTime = 1; (* assuming a maximum distance of 200 meters *)
          TimeOUT = 300;           (* not prescribed by standard, might differ for ACK/CTS *)

          MACHeaderCRC = 272;      (* in bits; = MAC Header and frame check sequence CRC *)
          ACK = 112;               (* in bits; = acknowledgment frame *)
          RTS = 160;               (* in bits; = Ready-To-Send frame  *)
          CTS = 112;               (* in bits; = Clear-To-Send frame  *)

(* PHY-dependent parameters (DSSS) *)
parameter aRxTxTurnaroundTime=4; (* time to change from receive to transmit, <= 5 *)
          aCCATime=14;           (* time receiver needs to assess medium, <= 15 *)
          aSlotTime=20;          (* > vulnerablePeriod *)
          SIFS=10;               (* short interframe space, value fixed for DSSS *)
          DIFS=50;               (* DCF interframe space, value fixed for DSSS *)
          aCWmin=31;             (* initial value to determine contention window *)
          aCWmax=1023; bcmax=5;  (* related by aCWmax=2^(bcmax-1)*(aCWmin+1)-1 *)

          PLCPHeader = 48;            (* in bits *)
          PLCPPreamble=144/ShortFactor;  (* in bits *)

(* transmission times *)
parameter PHYHeaderTransTime = PLCPPreamble/MinBitRate + PLCPHeader/(ShortFactor*MinBitRate);
          MACHeaderCRCTransTime = MACHeaderCRC/B;
          ACKTransTime = ACK/B;
          RTSTransTime = RTS/B;
          CTSTransTime = CTS/B;

(* transmission times in case of success/collision *)
parameter vulnerablePeriod = aAirPropagationTime+aCCATime+aRxTxTurnaroundTime;
          if (Handshake=1) then       (* Basic Access *)_
            collConstTransTime = PHYHeaderTransTime+MACHeaderCRCTransTime+DIFS-aCCATime;
            collMaxBodyTransTime = MaxFrameBody/B;
            succConstTransTime = PHYHeaderTransTime+MACHeaderCRCTransTime+SIFS
                    +aAirPropagationTime+PHYHeaderTransTime+ACKTransTime+DIFS-aCCATime;
          else                        (* RTS/CTS *)
            collConstTransTime = PHYHeaderTransTime+RTSTransTime+DIFS-aCCATime;
            collMaxBodyTransTime = 0;
            succConstTransTime = 4*PHYHeaderTransTime+RTSTransTime+CTSTransTime
                +MACHeaderTransTime+ACKTransTime+3*SIFS+3*aAirPropagationTime+DIFS-aCCATime;

          succMaxBodyTransTime = MaxFrameBody/B;

end DSSSpar.
```

Figure 1.   SPNL module for the parameters of a DSSS WLAN

arrangement of summands reflects the composition of the various frames and handshakes.

2. The SPN model

By virtue of the hierarchical concept of SPNL [5] akin to Modula-2 or Ada, we compose the WLAN model of three modules (as adapted from [6]) from bottom to top: the parameter module DSSSpar already presented in Fig. 1 and the station module WLANstation, whose processtype *Station* implements the DCF (see Fig. 2) and is instantiated once for every station in the top-level module. Besides global measure definitions, this module – not shown in this paper – basically contains the two places vuln and busy connected to the process instances of type *Station* by arcs and ports (see port elements toVuln, fromVuln, toBusy, fromBusy). Similarly, the rate rewards (SPNL keyword **rreward**) Nvuln and Nbusy denote the corresponding numbers of tokens in these places, respectively. Due to limited space, we must assume the reader's familiarity with stochastic Petri nets and SPNL. Besides conventional SPN elements, we make use of transitions with general firing time distributions (deterministic, (discrete)uniform, see **dist**), different firing policies (**prs** = preemptive resume, **prd** = preemptive repeat different as default, see **policy**), guards (keyword **guard**) and marking-dependent arc multiplicities (e.g., #bc in Fig. 2).

We now point out how the core of the SPNL model (see Fig. 2) reflects the behavior of the DCF described in Sec. 1.1. Place free models the free buffer places – initially K – for frames in the MAC layer of a station. Exponential transition gen models the arrival of data units from higher protocol levels and place wait models buffered frames waiting for transmission. The remaining part of the SPN represents the channel access and transmission of frames.

Tokens in wait may enter the lower subnet, when a token is in place idle or fin – either via immediate transition first (to place sense) or via consecutive (3(b) in Sec. 1.1). In place sense (sensing of channel), the corresponding guards (see transatt in Fig. 2) cause either of immediate transitions defer and access to fire thus starting the backoff procedure in case of a busy medium (token to place Pbackoff, 3(a)) or the transmission in its vulnerable period (token to place Pvuln, 2 in Sec. 1.1). Tokens in place busy represent the number of perceptible ongoing transmissions. If in the vulnerable period any other station accesses the medium, a collision occurs. The number of transmissions in the vulnerable period is represented by tokens in place vuln (on top level). A collision corresponds to more than one token in vuln. In this case immediate transition Tcoll fires and puts a token in place Pcoll, which serves as a "model artifact"-memory that a collision has occurred.

```
module WLANstation;
  use DSSSpar;

  processtype Station(param K, lambda; rreward Nvuln, Nbusy);
  port toVuln, toBusy: t->p;
       fromVuln, fromBusy: t<-p;
  smeasure throughput, waitingTime;
private
```

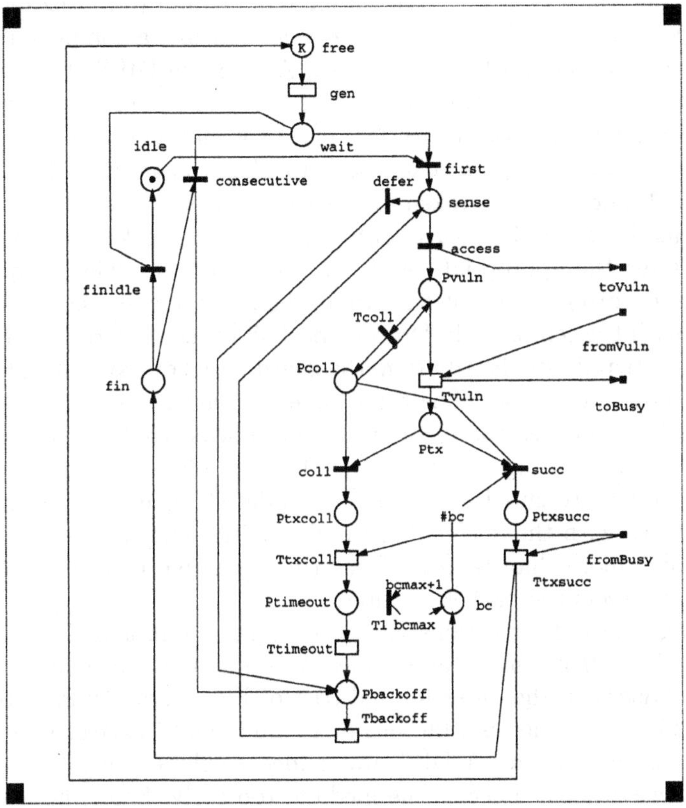

```
smeasure throughput = (E{#gen}*L)/B;
         waitingTime = (K- E{#free})/ E{#gen};
transatt gen: dist = exp(lambda);
         Tvuln: dist = det(vulnerablePeriod);
         Ttxcoll: dist = uniform(collConstTransTime, collConstTransTime+collMaxBodyTransTime);
         Ttxsucc: dist = uniform(succConstTransTime, succConstTransTime+succMaxBodyTransTime);
         Ttimeout: dist = det(TimeOut-DIFS);
         Tbackoff: dist = discreteuniform(0,((2^#bc)*(aCWmin+1)-1)*aSlotTime, aSlotTime),
                   guard = Nbusy=0, policy = prs;
         defer: guard = Nbusy > 0;
         access: guard = Nbusy = 0;
         Tcoll: guard = Nvuln > 1;
  end Station;
end WLANstation.
```

Figure 2. The core of the DCF

The duration of the vulnerable period is represented by the deterministic transition Tvuln.

Depending on whether a token is in Pcoll or not, either the immediate transition coll (enabling Ttxcoll) or succ (enabling Ttxsucc) fires. The firing time of transition Ttxsucc reflects the duration of the successful frame exchange sequence of the employed technique (either BA or RTS/CTS). The firing time of transition Ttxcoll takes into account only the frame, which contended for the medium (data packet or RTS, respectively). In both cases aCCATime, which is a component of the vulnerable period, must be subtracted. Appending DIFS at the end of the transmission times simplifies the modeling of channel sensing (1 and 2 in Sec. 1.1): otherwise transition access would have to be timed. The firing of Ttxsucc puts a token back to places free and fin, and the firing of Ttxcoll puts a token into place Ptimeout.

Deterministic transition Ttimeout forwards the token to place Pbackoff after the appropriate ACK or CTS timeout. The complex guard and the firing policy of transition Tbackoff model the backoff procedure with the temporarily "frozen" backoff timer. Before sampling from its discrete uniform distribution, the guard based on "system" place busy must evaluate to $true$. The firing time of Tbackoff is sampled or decremented, as soon as the medium has been sensed idle for a DIFS (#busy = 0). After firing, a token is put back to place sense and the backoff counter represented by place bc is updated. The self-loop from place bc via T1 with the depicted arc multiplicities guarantees that bc never contains more than $bcmax$ tokens (in a tangible marking). The firing of succ resets the backoff counter.

The firing times of transitions gen, Tvuln, Ttimeout and Tbackoff are independent of the chosen access mechanism and are given after the keyword **transatt** in the lower textual part of Fig. 2 (with parameters defined in Fig. 1.) The distributions of Ttxsucc and Ttxcoll depend on the access mechanism. A DIFS was added to these firing times to prevent other stations to access the medium too early. Since, however, in case of a collision, the sending station starts its timeout counter right after the end of its transmission, this DIFS has to be subtracted again from $TimeOut$ in the delay of Ttimeout. The payload $FrameBody$ may be arbitrarily distributed over $[0, MaxFrameBody]$ with a mean length equal to L. We chose a uniform distribution over the whole interval.

Stationary performance measures can easily be expressed in terms of rate and impulse rewards. $E\{\#P\}$ gives the expected number of tokens in place P and $E\{\#T\}$ the mean number of firings per unit time of transition T. The measures throughput and mean waiting time are defined for each station in processtype $Station$ right below the graphical area (keyword **smeasure** in Fig. 2). The throughput is normalized to bit rate B. The mean waiting time accounts for the time from data generation until the end of transmission and is obtained by Little's law. For system indices, (station) throughputs are added and mean waiting times are averaged. The model parameter λ (see transition **gen**, written out as *lambda* in the figures) is determined from the input parameter V, the "virtual load", by $V = N\lambda L/B$. V is the load, if the buffer had infinite capacity $K = \infty$ (i.e., the arrival process is not interrupted), normalized to bit rate B.

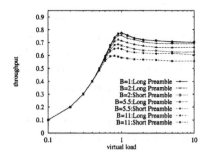

Figure 3a.   Basic Access: S vs. V

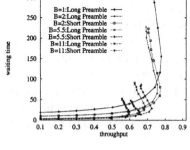

Figure 3b.   Basic Access: W vs. S

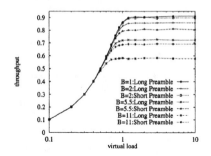

Figure 3c.   RTS/CTS: S vs. V

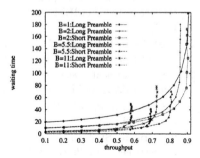

Figure 3d.   RTS/CTS: W vs. S

3.   Simulation results

The results of this section were computed by means of the SPNL simulation component of the software tool TimeNET [9] with a confidence level of 99% and a maximum relative error margin of 1%. This yielded confidence intervals too small for a reasonable representation in the figures. Except for parameters B, *ShortFactor* and *Handshake*, which define the transmission mode uniformly adopted by the ten stations in the WLAN, and for the varying virtual load V, the DSSS parameters are always set as in Fig. 1. All data packets are directed and need to be acknowledged.

For each of the 14 transmission modes, a simulation experiment yields two curves: (a) the throughput S versus the virtual load V (in a log-linear plot) and (b) the mean waiting time W (in ms) versus the throughput S (i.e., the delay-throughput characteristics). In Figs. 3a&b and 3c&d, we show results for the access mechanisms BA and RTS/CTS, respectively. Clearly, we observe the throughput deterioration of BA for high virtual loads in all cases, but less pronounced for higher bit rates. Especially in heavy-load conditions, the overhead introduced by the virtual collision detection via RTS/CTS pays off in terms of higher throughputs and shorter mean waiting times. Due to this overhead, the performance gain due to short preambles becomes more noticeable in RTS/CTS. Note that smaller average payloads would emphasize the effect of short preambles even more. Naturally, the normalized throughputs decrease with higher bit rates (since PHY headers are always sent at 1 or 2 Mbps), while mean waiting times are substantially reduced. Yet better performance is achieved, if an improved PHY reduces the vulnerable period compared to *aSlotTime* (in our experiments 19 vs. 20).

References

[1] H. S. Chaya and S. Gupta. Performance modeling of the asynchronous data transfer methods of IEEE 802.11 MAC protocol. *Wireless Networks*, 3:217–234, 1997.

[2] B. P. Crow, I. Widjaja, J. G. Kim, P. Sakai. IEEE 802.11 wireless local area networks. *IEEE Communication Magazine*, 116-126, 1997

[3] I. S. Department. *802.11: IEEE Standard for Wireless LAN Medium Access Control (MAC) and Physical Layer (PHY) Specifications*, 1999 Edition.

[4] I. S. Department. *802.11b: Higher Speed Physical Layer Extension in the 2.4 GHz Band*, 1999 Edition.

[5] R. German. SPNL: Processes as language oriented building blocks of stochastic petri nets. In *Proc. 9th Int. Conf. on Modelling Techniques and Tools for Computer Performance Evaluation*, LNCS 1469, pages 123–134. Springer, 1997.

[6] A. Heindl and R. German. Performance modeling of IEEE 802.11 wireless LANs with stochastic Petri nets. *Performance Evaluation*, 44:139–164, 2001.

[7] A. Köpsel, J. P. Ebert, and A. Wolisz. A performance comparison of Point and Distributed Coordination Function of an IEEE 802.11 WLAN in the presence of real-time requirements. *Proc. 7th Intl. Workshop on Mobile Multimedia Communications*, Tokio, Japan, 2000.

[8] J. Weinmiller, M. Schlager, A. Festag, and A. Wolisz. Performance study of access control in wireless LANs IEEE 802.11 DFWMAC and ETSI RES 10 HIPERLAN. *Mobile Networks and Applications*, 2:55–67, 1997.

[9] A. Zimmermann, J. Freiheit, R. German, and G. Hommel. Petri net modelling and performability evaluation with TimeNET 3.0. In *Proc. 11th Int. Conf. on Modelling Techniques and Tools for Computer Performance Evaluation*, Chicago, USA, 2000.

BUFFER DESIGN IN DELTA NETWORKS

Dietmar Tutsch
Technische Universität Berlin
Real-Time Systems and Robotics
D-10587 Berlin, Germany
DietmarT@cs.tu-berlin.de

Abstract   Multistage interconnection networks (MINs) are frequently proposed as connections in multiprocessor systems or network switches. Delta networks are a subset of MINs and are investigated in this paper. A generator is used to set up automatically systems of equations modeling such delta networks. The model allows to calculate analytically the performance of MINs, e.g. for various buffer sizes. Comparing different buffer sizes helps to determine the optimal size for a given traffic pattern including multicast traffic. The influence of the buffer size on the network throughput and delay for small and large switching elements is also examined. The optimal network parameter choice for a given traffic pattern can be found.

Keywords: analytical model, buffer size, multicasting, multistage interconnection networks

Introduction

Multistage interconnection networks (MIN) with the banyan and delta property are proposed to connect a large number of processors to establish a multiprocessor system as shown in Abandah and Davidson, 1996. They are also used as interconnection networks in Gigabit Ethernet (see Yu, 1998) and ATM switches (see Awdeh and Mouftah, 1995). Such systems require high performance of the network. To increase the performance of a MIN, Dias and Jump, 1981 inserted a buffer at each input of the switching elements (SE) and developed an analytical model to predict its performance. Buffers at each SE allow to store the packets of a message until they can be forwarded to the next stage in the network. In their model, Dias and Jump reduced each stage in the network to one SE of this stage so that it could be mapped onto a Markov chain.

G. Hommel and S. Huanye (eds.), The Internet Challenge: Technology and Applications, 93–101.
© 2002 *Kluwer Academic Publishers.*

Jenq, 1983 introduced a model with lower complexity than that of Dias and Jump by considering only one input port of an SE per stage to model the complete stage. Yoon et al., 1990 extended Jenq's model by using arbitrary buffer lengths in the network and arbitrary SE sizes. Theimer et al., 1991 proposed another extension of Jenq's model. They took into account the dependence between the packets of two successive time cycles in a single buffered MIN. Mun and Youn, 1994 combined this dependence with arbitrary buffer length. The influence of the network stage number on the clock cycle was investigated by Ding and Bhuyan, 1994. Atiquzzaman and Akhtar, 1995 and Zhou and Atiquzzaman, 1996 examined nonuniform traffic like hot spot traffic. Cut-through switching was taken into account by Widjaja et al., 1993, and by Boura and Das, 1997. On the other hand, there are a few investigations on multicast routing in MINs (see Park and Yoon, 1998; Ren et al., 1998; Sivaram et al., 1998) and on the structure of multicast ATM switches (see Guo and Chang, 1998; Sharma, 1999). An analysis of multicasting in MINs is presented by Yang, 1999. But in contrast to the other models, this model is not able to deal with the backpressure mechanism to handle full buffers.

Tutsch and Hommel, 1998 extended Jenq's model such that the analytical model additionally copes with performance analysis of a network with multicasting. Multicasting includes the two special cases of unicasting and broadcasting of messages. The model uses store and forward routing and the backpressure mechanism. Because an arbitrary finite buffer size in each stage is considered and the dependence between the packets of two successive time cycles are taken into account, the model is more accurate and powerful than in Tutsch and Holl-Biniasz, 1997.

Tutsch and Hommel modified their paper (see Tutsch et al., 2000) to enable performance analysis in case of cut-through switching. That allowed a comparison between the routing schemes of store and forward routing and cut-through switching.

In Tutsch and Hommel, 1998 and Tutsch et al., 2000, a system of equations was set up manually for performance estimation. During the set up some rules emerged to build such a system. These rules were extended for automatic generation of systems of equations in Tutsch and Hommel, 2002, which cope with the multicast performance analysis of MINs consisting of switching elements larger than 2×2. In this paper, an extensive investigation of various various buffer sizes benefits from the automatically generated equations. New insights to the influence of various network structures on the performance are determined.

The paper is organized as follows. The structure of the investigated MINs, the model and its assumptions are presented in Section 1. New analytical results concerning the network performance for various buffer sizes are derived in Section 2. Finally, Section 3 summarizes and gives conclusions.

1. Behavior and Structure of Delta Networks

The following results are received by performance evaluation of internally clocked $N \times N$ MINs consisting of $c \times c$ switches with $n = \log_c N$ stages (Figure 1). Such networks are called delta networks if there is one

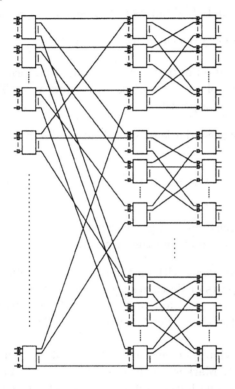

Figure 1.   3-stage delta network consisting of $c \times c$ SEs

unique way from a certain network input to a certain output (banyan network). It is additionally required that packets can use the same routing tag to reach a certain network output independently of the input at which they enter the network.

To achieve synchronously operating switches, the network is internally clocked. In each stage k $(0 \leq k \leq n - 1)$, there is a FIFO buffer of size $m_{max}(k)$ in front of each switch input. The packets are routed by store and forward routing from a stage to its succeeding one by backpressure mechanism. Multicasting is performed by copying the packets within the $c \times c$ switches.

The assumptions for the model are the same as those of Jenq, 1983 (Model I). Most analytical MIN models deal with those assumptions to reduce the problem complexity in such way that an analytical model can be established: Uniform traffic and the independence of the packets are assumed and therefore, the reduction of each stage to one switch input and its buffer is allowed:

- The traffic load of all inputs of the network is equal.

- All packets have the same size (like in ATM).

- Their destination outputs are distributed uniformly. That means every output of the network is with equal probability one of the destinations of a packet.

- Conflicts between packets are solved randomly with equal probabilities.

- Packets are removed from their destinations immediately after arrival.

- Routing is performed in pipeline manner. That means the routing process occurs in every stage in parallel.

The $c \times c$ switches operate with partial multicasting: If anyone of the destination output ports or destination buffers is not available, the packet will stay in the present stage and copies will only be transmitted to the available destinations. The transmission to the other destinations is performed later.

Taking the previously mentioned behavior into account, an analytical model based on a Markov chain was set up. The state space of the model, which copes with $c \times c$ switches, represents all reachable states of the first buffer positions of the switches and the queue length within the buffers.

To calculate the steady state performance measures of the network, fixed point iteration is used. The iteration is initialized with the unloaded network. Then, packets are pushed into the network and in each iteration step, the state probabilities are determined depending on the probabilities of the previous step.

A detailed description of the state transitions, the automatic generation of the systems of equations, and the iterative algorithm can be found in Tutsch and Hommel, 2002.

2. MIN Performance

The automatic generation allows performance analysis of delta networks with arbitrary switch sizes c. The performance is represented by the normalized throughput S_i at the inputs of the network and the normalized throughput S_o at the outputs of the network, the mean delay time $d(k)$ of the packets in each stage, the mean delay time d_{tot} of the packets in the whole network, and the mean queue length $\bar{m}(k)$ of the buffers in each stage. The following figures show some results.

As multicasting is considered in this paper, many different assumptions about the shape of the network traffic are possible (see Tutsch and Brenner, 2000). The most simple case is to assume that all possible combinations of destination addresses for each packet entering the network are equally distributed. This traffic pattern is used in this paper.

Figure 2 investigates the buffer sizes of various network sizes. The networks are composed of 4×4 switches. The input traffic is offered with a load of 0.1. The buffer size is increased from $m_{max}(k) = 1$ to

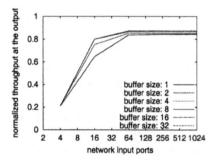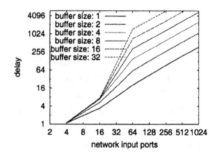

Figure 2. Comparing buffer sizes dependent on network size

$m_{max}(k) = 32$. The left hand figure shows an almost constant throughput rise by doubling the buffer size up to a size of $m_{max}(k) = 8$. A comparison of the delay times points out that larger buffer sizes cause higher delay times. The reason is obvious: Because of network congestion (almost all buffers are occupied) the packets have to traverse more buffer positions till they reach the first one in a stage. Being forwarded one position takes at least one clock cycle. If no congestion arises the buffer queue length is almost identical for different buffer sizes and delay times are equal. The networks are not congested up to a network size larger than $N = 16$ to $N = 64$ due to the large buffers and the light offered network load.

Figure 3 compares various buffer sizes changing the offered load to the network. The throughput and the delay of 64×64 MINs is shown. The network consists of 4×4 switches. The offered load of the network

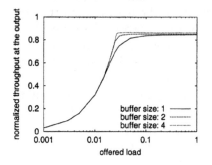 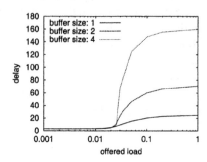

Figure 3. Comparing buffer sizes (4×4 switches)

is increased from 0.001 to 1. Using a light network load (up to 0.015), the network shows identical behavior concerning throughput and delay. That is because never more than one packet has to be stored in a buffer.

Increasing the load increases the amount of packets and thus, the number of conflicts at the switches: more packets occur that must be stored. Larger buffers allow a better performance because switches can store newly received packets even if the packet at the first buffer position is blocked several times.

An offered load higher than 0.1 leads to a congested network: all buffers are mostly occupied. Even larger buffers do not increase essentially the throughput. But the delay times explode due to long queues in the buffers at each network stage.

Figure 4 compares various buffer sizes for a switch size of 2×2. All other constraints are equal to the previous case. The relations of throughput and delay between the investigated buffer sizes are mostly similar to 4×4 switches. Nevertheless, there are some differences. If the network is congested the throughput in case of 2×2 switches is less than in case of 4×4 switches, especially if the buffer size is small. The delay times in case of 2×2 switches are higher. It originates from the larger number of stages. 64×64 MINs of 2×2 switches result in 6 stages, whereas networks of the same size but 4×4 switches result in 3 stages. More stages means more queues to pass and therefore, a higher consumption of time.

Concerning delay times, an offered load of around 0.02 leads to the effect of a lower delay time in case of larger buffer sizes. That is in opposition to any other load. An offered load of around 0.02 represents the transition from a network without any congestion to a congested

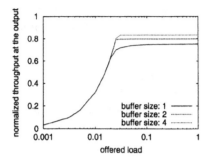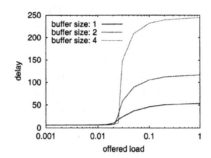

Figure 4. Comparing buffer sizes (2×2 switches)

network. Looking at the average queue lengths of each stage (not given in the figures) in this case shows a queue length of less than 1 for all stages. The stage delay times are rising from 1.1 (first stage, 2×2 switches) to 3.3 (last stage) in case of a buffer size of 4. But the stage delay times are falling from 4.3 (first stage, 2×2 switches) to 1.3 (last stage) in case of a buffer size of 1.

In other words: Due to an offered load that leads to a non-negligible amount of blockings but too less blockings to establish network congestion, packets must often stay in their buffer for several clock cycles. But in average, they could be forwarded in a higher rate than new packets are received by this stage: no congestion occurs. Nevertheless, it may occur that new packets are received while the old one is still in the buffer. If the buffer size is larger than $m_{max}(k) = 1$ (e.g. $m_{max}(k) = 4$), those new packets can be stored. If the buffer size is just $m_{max}(k) = 1$, the backpressure mechanism keeps those packets in the previous stage and packets destined to this previous stage are also kept in their previous stage, and so on. Forwarding such a packet, which is blocked via several stages, takes more time than forwarding a packet at the end of a queue: in the first case, the observed packet may be in conflict at each stage it must cross (and therefore delayed). In the second case, the observed packet may only be in conflict when it reaches the first buffer position.

3. Conclusions

With an analytical model, a much faster performance evaluation of buffered multistage interconnection networks is possible than using simulation. This paper benefits from such an analytical model. Various buffer sizes and their effect on the performance of delta networks is investigated. The influence of the switching element size and the offered load is also considered. The network traffic includes multicast traffic as well as unicast traffic. Performance is measured in terms of throughput and delay time.

The results show a different influence of the buffer size in case of network congestion and in case of a slightly loaded network. Concerning the delay time, some surprising effects are achieved for an offered load that results in a network that is in-between a congested and a non-congested state.

References

Abandah, G. A. and Davidson, E. S. (1996). Modeling the communication performance of the IBM SP2. In *Proceedings of the 10th International Parallel Processing Symposium (IPPS'96); Hawaii.* IEEE Computer Society Press.

Atiquzzaman, M. and Akhtar, M. S. (1995). Performance of buffered multistage interconnection networks in a nonuniform traffic environment. *Journal of Parallel and Distributed Computing*, 30(1):52–63.

Awdeh, R. Y. and Mouftah, H. T. (1995). Survey of ATM switch architectures. *Computer Networks and ISDN Systems*, 27:1567–1613.

Boura, Y. M. and Das, C. R. (1997). Performance analysis of buffering schemes in wormhole routers. *IEEE Transactions on Computers*, 46(6):687–694.

Dias, D. M. and Jump, J. R. (1981). Analysis and simulation of buffered delta networks. *IEEE Transactions on Computers*, C–30(4):273–282.

Ding, J. and Bhuyan, L. N. (1994). Finite buffer analysis of multistage interconnection networks. *IEEE Transactions on Computers*, 43(2):243–247.

Guo, M.-H. and Chang, R.-S. (1998). Multicast ATM switches: Survey and performance evaluation. *ACM Sigcomm: Computer Communication Review*, 28(2):98–131.

Jenq, Y.-C. (1983). Performance analysis of a packet switch based on single–buffered banyan network. *IEEE Journal on Selected Areas in Communications*, SAC–1(6): 1014–1021.

Mun, Y. and Youn, H. Y. (1994). Performance analysis of finite buffered multistage interconnection networks. *IEEE Transactions on Computers*, 43(2):153–161.

Park, J. and Yoon, H. (1998). Cost-effective algorithms for multicast connection in ATM switches based on self-routing multistage networks. *Computer Communications*, 21:54–64.

Ren, W., Siu, K.-Y., Suzuki, H., and Shinohara, M. (1998). Multipoint-to-multipoint ABR service in ATM. *Computer Networks and ISDN Systems*, 30:1793–1810.

Sharma, N. K. (1999). Review of recent shared memory based ATM switches. *Computer Communications*, 22:297–316.

Sivaram, R., Panda, D. K., and Stunkel, C. B. (1998). Efficient broadcast and multicast on multistage interconnection networks using multiport encoding. *IEEE Transaction on Parallel and Distributed Systems*, 9(10):1004–1028.

Theimer, T. H., Rathgeb, E. P., and Huber, M. N. (1991). Performance analysis of buffered banyan networks. *IEEE Transactions on Communications*, 39(2):269–277.

Tutsch, D. and Brenner, M. (2000). Multicast probabilities of multistage interconnection networks. In *Proceedings of the 12th European Simulation Symposium 2000 (ESS'00); Hamburg*, pages 554–558. SCS.

Tutsch, D., Brenner, M., and Hommel, G. (2000). Performance analysis of multistage interconnection networks in case of cut-through switching and multicasting. In *Proceedings of the High Performance Computing Symposium 2000 (HPC 2000); Washington DC*, pages 377–382. SCS.

Tutsch, D. and Holl-Biniasz, R. (1997). Performance evaluation using measure dependent transitions in Petri nets. In *Proceedings of the Fifth International Symposium on Modeling, Analysis and Simulation of Computer and Telecommunication Systems (MASCOTS'97); Haifa*, pages 235–240. IEEE Computer Society Press.

Tutsch, D. and Hommel, G. (1998). Multicasting in buffered multistage interconnection networks: an analytical algorithm. In *12th European Simulation Multiconference: Simulation – Past, Present and Future (ESM'98); Manchester*, pages 736–740. SCS.

Tutsch, D. and Hommel, G. (2002). Generating systems of equations for performance evaluation of buffered multistage interconnection networks. *Journal of Parallel and Distributed Computing*, 62(2):228–240.

Widjaja, I., Leon-Garcia, A., and Mouftah, H. (1993). The effect of cut-through switching on the performance of buffered banyan networks. *Computer Networks and ISDN Systems*, 26:139–159.

Yang, Y. (1999). An analytical model for the performance of buffered multicast banyan networks. *Computer Communications*, 22:598–607.

Yoon, H., Lee, K. Y., and Liu, M. T. (1990). Performance analysis of multibuffered packet–switching networks in multiprocessor systems. *IEEE Transactions on Computers*, 39(3):319–327.

Yu, B. (1998). Analysis of a dual-receiver node with high fault tolerance for ultrafast OTDM packet-switched shuffle networks. Technical paper, 3COM.

Zhou, B. and Atiquzzaman, M. (1996). Efficient analysis of multistage interconnection networks using finite output-buffered switching elements. *Computer Networks and ISDN Systems*, 28:1809–1829.

ARCHITECTURES FOR ADVANCE
RESERVATIONS IN THE INTERNET

L.-O. Burchard

Telecommunication Systems Institute
Technical University of Berlin
baron@cs.tu-berlin.de

Abstract   Some applications, mainly grid computing or video conferencing require
the transmission of large amounts of data or the reservation of a certain
amount of bandwidth on the network. Usually the timing parameters
such as start and duration or total amount of data for these transmission
is known in advance. However, todays networks do not provide means
for enabling advance reservations. In this paper, we present two possible architectures for implementing advance reservation mechanisms on
IP networks. The architectures are based on integrated and differentiated services networks. We discuss the advantages and disadvantages
of both architectures and how advance reservation mechanisms can be
implemented in the respective environment.

Keywords: Network QoS, Advance Reservations, Architecture

1.   Introduction

In order to enforce quality-of-service (QoS) guarantees on the network,
generally two types of technologies are available. The integrated services
approach (IntServ) uses the RSVP protocol in order to establish per-
flow QoS guarantees while the *differentiated services* (DiffServ) model
relies on few service classes with different priorities. DiffServ requires an
additional management component in order to facilitate end-to-end QoS
guarantees. *Bandwidth brokers* are considered as a managing system in
such networks: a client (person or software) sends a reservation request
to the broker, which then grants or denies access to the network. A
single broker is only responsible for managing a certain part of a network
(*domain*), e.g. a corporate network or the network of an ISP.

Using these technologies, two types of bandwidth reservations can be
considered: *immediate* and *advance reservations*. In contrast to immediate reservations, where the network reservations are established im-

G. Hommel and S. Huanye (eds.), The Internet Challenge: Technology and Applications, 103–110.
© *2002 Kluwer Academic Publishers.*

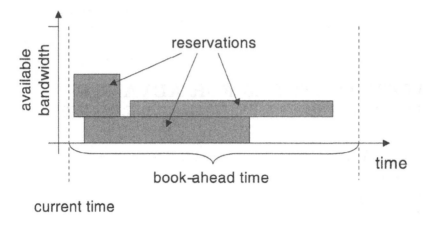

Figure 1.   Advance reservations and book-ahead

mediately after the request is admitted, advance reservations allow to specify and request a given QoS for a reservation a long time before the actual transmission is to be made. In the following, the time for which reservation requests can be made is called *book-ahead time* (see Figure 1).

Advance reservations are especially useful in any environment where large amounts of data have to be transferred over a network as is the case in grid computing where processes on different parallel computers must communicate with each other. Another example are web caches or distributed media servers which must be filled with large amounts of content (e.g. media files). Especially in cases where long-lasting transmissions of large amounts of data occur (usually in those cases transmission times are know in advance) an advance reservation mechanism provides a useful tool for guaranteeing the required network resources.

One important task to be performed in such an environment is *admission control*, which means reservation requests have to be checked whether sufficient resources are available for the duration of the reservation. In the scenario examined in this paper, only advance reservations were considered. Admission control for immediate reservations is less complex, in particular it is just a special case without book-ahead time.

In addition to that, routing plays a major role during the admission control task. It is required to consider not only the way to find a path (e.g. using Dijkstra's Shortest Path algorithm) but also the time for the routing process has to be taken into account. The timing aspect adds a new dimension to the routing procedure and therefore results in a much more complex task.

In this paper, we present and compare two architectures for advance reservation mechanisms based on common QoS mechanisms for the Internet. Both integrated services and differentiated services will be examined and their advantages and disadvantages will be presented.

In the following sections, after discussing some related work, the two architectures are presented and evaluated. The paper is concluded with some final remarks.

2.   Related Work

Quality-of-service (QoS) in networks has been addressed for a variety of reasons. For some applications such as video and audio streaming it is required in order to avoid network jitter and congestion. Moreover, when it is necessary to deliver a large amount of data over a network within a given time, it is required to reserve network resources in order to meet the given deadline. This is the case in grid computing environments and also distributed multimedia applications (Burchard and Lüling, 2000).

The requirement for advance reservations has been identified in several papers (Degermark et al., 1995; Ferrari et al., 1995). In addition to that, the relation between advance reservations on networks and advance reservations at the end-points of the network traffic, i.e. the computers involved, is described by (Wolf and Steinmetz, 1997). Those early works mainly concentrate on design issues for enabling advance reservations and present extensions to existing network reservation protocols such as RSVP (Schill et al., 1998).

In (Schelen and Pink, 1997), an architecture for advance reservations based on OSPF was proposed, using information from the routing tables of each router in order to determine routes for advance reservations. This approach heavily relies on how likely route changes with the bandwidth broker controlled network occur.

Most recent works deal with using differentiated services and introduce the concept of bandwidth brokers for managing network resources (Khalil and Braun, 2001; O. Schelen and S. Pink, 1998). In (Nichols et al., 1999) the basic principle of wide scale deployment of DiffServ using two service classes (premium and best-effort service) on the Internet using bandwidth brokers is described. Both works do not consider the usage of advance reservation. However, the approach using bandwidth brokers seems also to be suitable for advance reservations as presented in section 3.2.

Admission control in such an architecture has been studied by some papers, some of them using probabilistic approaches (Wischin and Greenberg, 1998) for assuring sufficient bandwidth is available. In (Burchard and Heiss, 2002), two data structures for admission control are compared with respect to their admission speed.

3. Architecture

Two possible architectures for a distributed advance reservation service will be presented and compared in this section.

3.1 Integrated Services

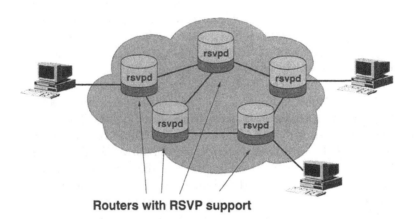

Routers with RSVP support

Figure 2. Architecture based on IntServ using routers with RSVP daemons

In general, resource reservation mechanisms in IP networks can be divided into two groups: integrated services (IntServ) and differentiated services (DiffServ) networks.

The resource reservation mechanism based on IntServ uses RSVP (ReSource ReserVation Protocol) as a signalling protocol. RSVP itself is rather unsuited for large-scale reservations because of its soft-state, per-flow-reservation approach. On each router, an RSVP daemon (rsvpd, see Figure 2) resides which is responsible for processing RSVP signalling messages and for the configuration of the router in order to reserve the respective share of the overall amount of bandwidth for a certain traffic flow.

Applying such a technology in the Internet results in an enormous amount of states to be kept on each router and therefore an RSVP based approach does not scale well. In order to implement advance reservation mechanisms based on RSVP, it is not only required to add extra timing information to the respective RSVP signalling messages (which require changes to the routers since the extra information is not defined in the RSVP standard, see (Schill et al., 1998)) but also to rely on the soft-state architecture. Moreover, admission control is an important issue for any RSVP based reservation architecture: it is required to restrict access to the network, i.e. access control mechanisms have to be enforce for the RSVP signalling. Although some approaches (Boyle et al., 2000) based on an additional protocol have been made, this makes the architecture more complicated.

Admission control in the IntServ scenario has to be performed independently on each router. This means for the requested period of time, each router must determine whether sufficient resources are available.

Routing in an IntServ environment is not an issue, since each router on the path determines independently the next hop and therefore no centralized approach is required in order to calculate paths through the network. The drawback of this solution is that routes may change between the time a request and the admission decision was made and the time the transmission actually starts. This is a critical point since it cannot be guaranteed that sufficient resources are available throughout the transmission. Although the RSVP protocol supports route changes by sending refresh messages along the reserved path during the time of the reservation, the availability of resources is not assured.

Summarizing, RSVP does not provide sufficient means for an advance reservation mechanism (extra timing information must be added and therefore the protocol has to be extended) and additionally does not scale well. Although admission control however can be performed in a decentralized way avoiding a single point-of-failure, route changes cannot be handled in an appropriate way.

3.2 Differentiated Services

In contrast to the IntServ based approach presented in the previous section, differentiated services are based on a *class-of-service* model rather than the per-flow approach. This means, reservations cannot be made for a single flow but for a distinct amount of bandwidth on each router. This amount of bandwidth is available for a certain class of flows, e.g. only for video transmissions. Such flows are marked using a special field in the IP header (DiffServ code point DSCP). This requires only to keep on each router the information how much bandwidth is available to each service class. In general, only a few service classes ((Nichols et al., 1999) propose two classes) are established which leads to a drastically simplification of the architecture compared to IntServ.

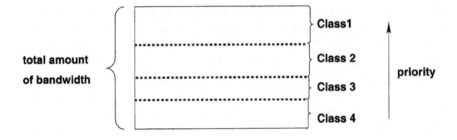

Figure 3. Service classes under DiffServ architecture with different amounts of bandwidth for each class

The DiffServ architecture however requires also access control features. In contrast to IntServ, it is sufficient to restrict access to the core network at the edge routers. This means, the DiffServ code point of each IP packet has to be altered on these routers according to the actual service class a particular traffic flow belongs to. In addition, in case such a traffic flow exceeds its bandwidth limits, *traffic shaping* has to be performed. However, these procedures only have to be applied on the edge routers, core routers do not have to deal with similar tasks. This allows to keep administration overhead low on the core routers. Consequently, such an architecture scales much better than IntServ based schemes. In typical networks, edge routers have to deal with a lot fewer traffic flows than core routers therefore the additional overhead for traffic shaping and admission control can be considered to be feasible.

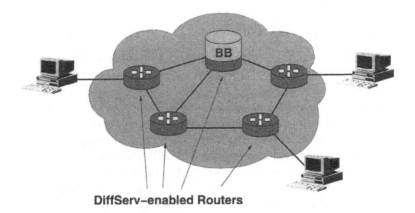

Figure 4. Architecture based on DiffServ

In a DiffServ environment, it is required to have an additional administration component (usually called *bandwidth broker*) which keeps track of the current network state in order to perform admission control, i.e. grants or denies access to the network. The drawback of this approach is that the bandwidth broker represents a single point-of-failure (which can be avoided e.g. by using multiple copies of the broker).

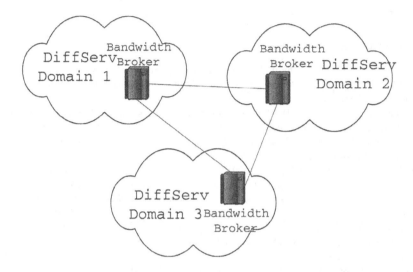

Figure 5.   Interaction of bandwidth brokers in several domains

Obviously, a single broker cannot be used to control an arbitrarily large network. Therefore, the concept of *domains* is introduced. A single broker is only responsible for a single domain, which represents a small part of e.g. the Internet. The network of an ISP or the network of a university can be examples of domains. In order to allow reservations across multiple domains, brokers in adjacent domains have to inter-operate in an appropriate way (see Figure 5).

4.   Conclusion

In this paper, we presented architectures for an advance reservation mechanism in IP networks. Two possible architectures have been presented, based on IntServ and DiffServ.

In order to offer new services such as advance reservations as required e.g. by grid computing environments or distributed media servers, the basic architecture has to be carefully designed. It is unlikely that other than the two approaches presented which are based on the current state-of-the-art technologies for enabling QoS in IP networks are applicable.

Since architectures based on IntServ do not scale well and in moreover require significant changes to existing protocols, DiffServ is preferred since it provides sufficient means for enabling advance reservations on a larger scale than IntServ. In order to achieve scalability in the DiffServ scenario, domains can be used which are each controlled by a single broker.

References

Boyle, J., Cohen, R., Durham, D., Herzog, S., Rajan, R., and Sastry, A. (2000). RFC2748: The COPS (Common Open Policy Service) Protocol. ftp://ftp.isi.edu/in-notes/rfc2748.txt.

Burchard, L.-O. and Heiss, H.-U. (2002). Performance Evaluation of Data Structures for Admission Control in Bandwidth Brokers. In *Proceedings of the 2002 Intl. Symposium on Performance Evaluation of Computer and Telecommunication Systems (SPECTS), to appear.*

Burchard, L.-O. and Lüling, R. (2000). An Architecture for a Scalable Video-on-Demand Server Network with Quality-of-Service Guarantees. In *Proceedings of the 5th Intl. Workshop on Distributed Multimedia Systems and Applications (IDMS), Lecture Notes in Computer Science, Springer,* volume 1905, pages 132–143.

Degermark, M., Kohler, T., Pink, S., and Schelen, O. (1995). Advance Reservations for Predictive Service. In *Network and Operating System Support for Digital Audio and Video,* pages 3–15.

Ferrari, D., Gupta, A., and Ventre, G. (1995). Distributed Advance Reservation of Real-Time Connections. In *Network and Operating System Support for Digital Audio and Video,* pages 16–27.

Khalil, I. and Braun, T. (2001). Implementation of a Bandwidth Broker for Dynamic End-to- End Resource Reservation in Outsourced Virtual Private Networks. In *25th Annual IEEE Conference on Local Computer Networks (LCN), November 9-10 2000,* pages 160–174.

Nichols, K., Jacobson, V., and Zhang, L. (1999). RFC2638: A Two-bit Differentiated Services Architecture for the Internet. ftp://ftp.isi.edu/in-notes/rfc2638.txt.

O. Schelen and S. Pink (1998). Resource Reservation Agents in the Internet. In *8th International Workshop on Network and Operating Systems Support for Digital Audio and Video (NOSSDAV'98).*

Schelen, O. and Pink, S. (1997). An Agent-based Architecture for Advance Reservations. In *22nd Annual Conference on Computer Networks (LCN'97).*

Schill, A., Breiter, F., and Kuhn, S. (1998). Design and evaluation of an advance reservation protocol on top of RSVP. In *4th International Conference Broadband Communications.*

Wischin, D. and Greenberg, A. (1998). Admission Control for Booking Ahead Shared Resources. In *Proceedings of IEEE INFOCOM 1998.*

Wolf, L. C. and Steinmetz, R. (1997). Concepts for Resource Reservation in Advance. *Multimedia Tools and Applications,* 4(3):255–278.

PERFORMANCE PREDICTION OF JAVA PROGRAM FOR METACOMPUTING

Martin Alt and Sergei Gorlatch
Technische Universität Berlin, Germany

Abstract   We address the challenging problem of program design for *metacomputing* in the heterogeneous, highly dynamic Internet environment, by providing the application user with a set of parameterized components called *skeletons*. This paper focuses on estimating the execution time of the application-specific program parts expressed in Java. Our results allow a reliable performance prediction for the execution of skeleton-based programs on remote Internet servers.

Keywords: programming models for metacomputing, performance prediction and evaluation, algorithm design, algorithmic skeletons, computational Grid.

1.   Introduction

Metacomputing aims at combining different kinds of computational resources connected by the Internet and making them easily available to a wide user community. Especially the emerging Java technologies enable shipping platform-independent Java bytecodes between different locations and their execution on remote machines. For time-consuming applications, it may well be the case that such remote computation on a high-performance server is faster and more cost-effective than when performed locally on the client. This idea leads to large-scale metacomputing systems, also known as *Computational Grids* (Foster and Kesselmann, 1998).

While the enabling infrastructure for metacomputing has made much progress recently, the design of distributed software remains a largely *ad hoc*, not well-understood activity. The main difficulty is the heterogeneous and dynamic nature of the target platforms, which leads to difficult-to-predict program behaviour. With many machines of different architectures involved and the communication taking place over a network of changing latency and bandwidth, it is fairly difficult to make systematic design decisions in the process of algorithm and program development. The resulting suboptimality of programs can hardly be compensated in the later implementation phases and can thus dramatically worsen both the speed and cost of computation.

G. Hommel and S. Huanye (eds.), The Internet Challenge: Technology and Applications, 111–119.
© 2002 *Kluwer Academic Publishers.*

In this paper, we address the performance predictability of metaprograms constructed using parameterized software components, so-called *skeletons*. In particular, we present a novel approach to estimating the performance of Java bytecodes, which are used in customizing skeletons for specific applications. Our goal is to enable a client to estimate in advance the expected runtime of a program constructed of skeletons on a remote server.

The remainder of the paper is structured as follows:

- We describe how metacomputing is organized using skeletons and introduce our approach to predicting program runtimes on a particular server, without actually involving the server (Section 2).

- We present a timing model for a remote server built by measuring a priori the runtime of automatically generated test programs (Section 3).

- We evaluate the quality of performance prediction by experimental measurements for two sample Java codes (Section 4)

We conclude the paper by discussing our results in the context of related work.

2. Programs using Skeletons

We consider programs that are built of high-level skeletons, available on some Internet servers. Skeletons are program building blocks that express typical parallel and distributed patterns of program behaviour. Skeletons are customizable by means of parameters, which are usually relatively simple Java codes capturing application-specific computations.

Figure 1 shows the code for matrix multiplication which can be used for customizing a skeleton. Class MatMult implements the interface CustOp, indicating that it provides a customization operator. The interface only contains the method apply, computing the resulting matrix from two input matrices.

```
class MatMult implements CustOp {
  public final static float matrixSize = 16;
  public  float[] apply(float[] a, float[] b) {
    out = new float[matrixSize];
    for (float i = 0; i < matrixSize; i++)
      for (float j = 0; j < matrixSize; j++)
        for (float k = 0; k < matrixSize; k++) {
          out[i*matrixSize+k] = a[i*matrixSize+j]*
            b[j*matrixSize+k];}
    return out;}
}
```

Figure 1.   Java code for the matrix multiplication operator used to customize a skeleton.

A simple but popular example of a skeleton is *reduction*, denoted $reduce(\otimes)$, where $\otimes$ is the customizing binary operator. Reduction performs the following computation: $reduce(\otimes)(x_1, \ldots, x_n) = (\ldots (x_1 \otimes x_2) \otimes \ldots \otimes x_n)$. For an associative $\otimes$, which is e.g. the case for matrix multiplication, the reduction skeleton can be implemented in parallel.

Since the set of skeletons is limited, servers can provide a reliable information about the nominal performance of their skeletons. For example, the runtime for reduction over an input array of length n on p processors working on shared memory is given by:

$$t_{red}(n, p) = (\lceil n/p \rceil - 1)t_{\otimes} + (p - 1)t_{\otimes}$$

where $t_{\otimes}$ is the run time of the customizing operator.

The problem we address in the remainder of the paper is how $t_{\otimes}$ can be predicted by the client for an arbitrary Internet server without actually involving that server. We have thus reduced the problem of performance prediction to the task of estimating the execution time for the – relatively small and simple – customizing operators, which are Java programs.

3. Timing Model

In the distributed setting, a desirable approach to server time estimation should not involve the server itself, because such involvement would impose a large time burden and jeopardize the whole enterprise. An intuitive approach to estimating the runtime of Java bytecodes consists of analyzing the code to determine how often each instruction is invoked. The obtained number and sequence of invocations can then be used to predict the runtime of the operator using a *timing model* for the server machine.

In this paper, our timing model is a vector with runtimes for each instruction type. Of course, timing models may also take cache and pipelining information into account to increase prediction accuracy (cf. Atanassov et al., 2001).

3.1 Combinations of Bytecode Instructions

A straightforward way of measuring the runtime for a particular bytecode instruction would be to execute a loop containing only instructions of that kind and divide the measured time by the number of iterations. However, since bytecode instructions are stack-oriented, every operand for an instruction must be brought to the stack before execution and then popped from the stack when the instruction is invoked. Therefore, only the runtime of instruction combinations — rather than of individual instructions — can be measured this way.

Our approach is to compute the runtimes of single instructions from the measured runtimes of relatively small test programs. For m different instruction kinds, each program or code sequence can be represented as an m-dimensional "program vector" $p = (p_1, p_2, \ldots, p_m)$ where the i-th entry indicates that instruction i occurs p_i times in the program.

The runtime for the program characterized by vector p can be written as

$$p \cdot t = \sum_i p_i t_i = T \tag{1}$$

with $\cdot$ denoting scalar product and t being the vector $(t_1, \ldots, t_m)$ containing the – unknown – runtimes t_i for instructions of type i. Time T is the execution time for the whole program, which can be measured.

If the runtime for n different programs is measured, then we obtain the following system of linear equations:

$$At = \begin{pmatrix} p_{(1)}^T \\ p_{(2)}^T \\ \vdots \\ p_{(m)}^T \end{pmatrix} \begin{pmatrix} t_1 \\ t_2 \\ \vdots \\ t_m \end{pmatrix} = \begin{pmatrix} T_1 \\ T_2 \\ \vdots \\ T_n \end{pmatrix} = \hat{T} \tag{2}$$

Here, matrix A contains the program vectors $p_{(i)}$ describing the n different programs, and $\hat{T}$ contains the measured time values T_i for each program. As before, t is the vector of time values for single instructions that are to be computed. This system of equations can be solved to obtain values for t.

If the number of test programs is greater than the number of instruction types, i. e. $n > m$, then the system of equations is overdetermined and a solution that fits all equations may not exist. Instead of solving $At = \hat{T}$, we then compute t that minimizes $\chi^2 = (\hat{T} - At)^2$. To solve such least-squares problems, we can use e. g. Singular Value Decomposition (Press et al., 1992).

The timing vector t is computed once for every server of a metacomputing system. On request, the timing vector is sent to a client, who can use it to estimate the runtime of a skeleton on that server.

3.2 Requirements for Test Programs

Now the question is how to construct the matrix A of (2), i. e. how large the test programs should be, and how many programs should be measured.

Size of Test Programs As it is not possible to measure the runtime of instructions directly, the alternative would be to measure programs consisting of two instruction types: an instruction that produces a stack element and one consuming an element. However, it appears that the JVM can extensively optimize such small programs, leading to unrealistic runtimes. For example, we measured the times for executing 1,000 integer additions and 1,000 integer multiplications in a loop of 10^5 iterations. Within the loop, for each `iadd`/`imul` instruction, one more instruction is necessary to bring the second operand to the stack.

Table 1. Left: Runtime for a loop containing addition, multiplication, a combination of both, and runtimes for loops containing two inhomogeneous code sequences ($\sim$ 1,000 instructions). *Right:* Comparison of runtime for 10^8 iterations of an empty loop with an empty vs. nonempty stack (one element), using SUN's JDK 1.3.1 on an UltraSparc III at 750 MHz.

	add	mul	add + mul	addmul
Time	286 ms	1429 ms	1715 ms	2106 ms

	P_1	P_2	$P_1 + P_2$	P_{12}
Time	1726 ms	1641 ms	3367 ms	3341 ms

	Empty Stack	Nonempty
Time	2564 ms	17400 ms

The values obtained are given in the "add" and "mul" columns of the first row of Table 1 (left). The time for executing a loop with both addition and multiplication (along with the indispensable stack-adjustment operations) would be expected to be the sum of the loops containing only addition or multiplication. In fact, an even shorter time could be expected because the combined loop contains more instructions per loop iteration, resulting in less overhead. However, the measured value ("addmul" in the first row of Table 1) of 2106 ms is considerably (approx. 23%) larger than the expected value of 1715 ms. Apparently, the JVM can excessively optimize loops that contain only arithmetic instructions of one type, but fails to do so for loops containing different instructions. Therefore, the `iadd` and `imul` loops are extensively optimized, while the combined loop is not, resulting in the longer execution time.

By contrast, when measuring inhomogeneous code sequences, linearity does hold. In the second row of Table 1 (left), the runtimes taken for two code sequences P_1 and P_2 of approx. 1,000 randomly chosen instructions are shown. One requirement for the construction of test programs is therefore that they should not be too small and homogeneous.

Number of Test Programs. Another question concerns the size of matrix *A* in (2). While the number of columns is given by the number of different instructions types, the number of rows equals the number of test programs measured. For the system to have a solution, the number of test programs must be at least equal to the number of instruction types. However, to avoid an excessive influence of measurement errors on the result, it is advantageous to measure many more programs than instructions.

Measuring more programs is also important because instructions of the same type may have different runtimes, depending on the execution context. For example, the time taken to execute a loop with an empty body varies considerably, depending on whether the JVM's stack is empty before loop execution. As shown by Table 1 (right), with a single element on the stack, the execution time is almost an order of magnitude larger compared to the time with an empty stack. This effect may be explained by the fact that data on the stack prevent the JVM from allocating the loop-counting variable (which is loaded to the stack and incremented in each loop iteration) to a register. Thus, a memory/cache access is necessary for every loop iteration. It is therefore advantageous to measure more programs to cope with such effects.

Experimentally, we have found that timing vectors converge to a mean value after measuring approximately 3,500 test programs. For fewer programs, the runtimes of single test programs still have a great influence on the computed timing vector.

3.3   Automatic Test Code Generation

As the previous consideration shows, obtaining satisfactory timing models requires, first, relatively large and inhomogeneous test programs, and second, a large number of test programs. Since it is practically impossible to produce a sufficiently large number of test programs "by hand", we generate them automatically, randomly selecting the bytecode instructions in them. We have implemented a bytecode generator to produce source files for the Jasmin bytecode assembler (Meyer and Downing, 1997). It generates bytecode containing randomly selected instructions, including simple loops. For more details on the generation algorithm, see (Alt et al., 2002).

The bytecode generator receives two parameters: the probability distribution of the instruction types and the length of code to be generated. For our initial experiments, we have simply used a uniform distribution of instruction types; we intend to investigate the effects of different distributions in the future. As for the length of test programs, the prediction was rather inaccurate for very small test programs of less than 500 instructions. On the other hand, if the test programs are too large, cache effects have a considerable influence on the measured runtimes. We therefore chose test programs with sizes varying between 500 and 4,500 instructions.

4.   Experimental Results on Performance Prediction

In this section, we first summarize our method of performance prediction, based on the criteria presented in the previous sections, and then present some experimental results from application of the method.

4.1   Performance Prediction Procedure

The procedure for estimating the runtime of a customizing operator (expressed as a Java program) consists of the following steps:

1 The execution time for a number of automatically generated test programs is measured. This results in a system of linear equations of the form (2).

A solution of the system is computed, resulting in a timing vector, which contains a runtime estimate for each instruction. An excerpt of the vector for the SunFire server is given in the second row of Table 2. The values were obtained by measuring the execution time for 4,000 automatically generated test programs.

2 On the client side, the customizing operator, e.g. the matrix multiplication listed in Figure 1, is implemented and compiled to Java Bytecode.

3 The operator is analyzed on the client to obtain the number of instruction invocations, i.e. the operator's program vector. A part of the vector for matrix multiplication is given in the third row of Table 2.

4 The client retrieves the server's timing model and computes the performance estimate for the operator by computing the scalar product of the program and the timing vectors, as in equation (1).

Table 2. Program vector for matrix multiplication and timing vector for an Ultra 5.

Instruction	...	iinc	bipush	fstore	fadd	fmul	...
Timing vector t [ns]	...	4.341	25.308	25.798	26.620	9.414	...
Program vector p	...	4368	16929	273	12288	16384	...

To validate our performance-prediction approach, we tested the prediction for two sequential Java codes: the first is the matrix multiplication operator listed in Fig. 1, the second is an operator from a tridiagonal system solver, described in more detail in (Bischof and Gorlatch, 2001) and (Alt et al., 2002). For the measurements, we used a Sun Ultra 5 Workstation with an UltraSparc IIi 360 MHz processor and a SunFire 6800 SMP machine with 16 UltraSparc III processors running at 750 MHz. The prediction results are given in the next paragraph.

Prediction Results.   The measured and predicted values are given in Table 3. Considering the relative errors, it becomes evident that the prediction results are better for the Ultra 5 than for the SunFire. This might, however, simply stem from the fact that the test programs for the former were measured on a dedicated machine, while there were always a number of other processes running on the latter (at least 4 out of 16 processors were used for other purposes).

Prediction Results.   The measured and predicted values are given in Table 3. Considering the relative errors, it becomes evident that the prediction results are better for the Ultra 5 than for the SunFire. This might, however, simply stem from the fact that the test programs for the former were measured on a dedicated

Table 3. Results for the prediction of two bytecodes: matrix multiplication (MM) and tridiagonal system solver (TDS).

	Ultra 5		SunFire	
	MM	TDS	MM	TDS
predicted time	492 μs	3.575 μs	241 μs	1.772 μs
measured time	556 μs	3.475 μs	213 μs	1.697 μs
absolute error	64 μs	0.100 μs	28 μs	0.075 μs
relative error	11.5 %	2.9 %	13.1 %	4.4 %

machine, while there were always a number of other processes running on the latter (at least 4 out of 16 processors were used for other purposes).

Considering that our timing models do not take into account any cache effects, the results are remarkably accurate. This is due to the fact that in both cases the operators and the data used during one operator invocation are small enough to fit into the processor's cache. For larger operators or data (e. g. matrices of larger size) cache effects would have to be taken into account.

5. Conclusion

We feel that algorithmic and programming-methodology aspects have been largely neglected in early metacomputing systems. Initial experience with computational grids has shown that entirely new approaches to software development and programming are required for such systems (Kennedy et al., 2002).

One of the main challenges in metacomputing is the unpredictable nature of the Internet resources, resulting in difficult-to-predict behaviour of algorithms and programs. While the runtime of metacomputing algorithms can be derived analytically, the corresponding expressions tend to be parameterized by the runtime of sequential parts of the algorithm. We propose predicting the runtime of the overall program from the runtime of these application-specific sequential parts, developing a novel performance-prediction method for Java programs.

Predicting the execution time of Java programs seems to have been largely neglected so far. Known approaches include using timing results of previous program executions, benchmarking (Au et al., 1996) or deriving estimates from the machine's peak performance (Berman et al., 2001). In the field of real-time systems, prediction of execution time for sequential code has been an active area of research for many years (e. g. Atanassov et al., 2001). However, the analysis of portable code, such as Java bytecode, has not been studied until recently. Initial research efforts (Bate et al., 2000; Bernat et al., 2000) are concerned with the high-level analysis of bytecode, i. e. the problem of counting how often an instruction is executed in the worst case. The concrete time values for single instructions are assumed to be given in the form of a "machine model".

In the paper, we presented a mechanism for computing run time values of metacomputing codes using automatically generated test programs. Based on the test measurements, our method allows us to predict the runtime of program executions. The application developer can thus obtain a concrete time value for the remote execution of the program, based on the sequential program parts.

References

Alt, M., Bischof, H., and Gorlatch, S. (2002). Program development for computational grids using skeletons and performance prediction. In *Third Int. Workshop on Constructive Methods for Parallel Programming (CMPP 2002)*, Technical Report. Technische Universität Berlin. To appear.

Atanassov, P., Puschner, P., and Kirner, R. (2001). Using real hardware to create an accurate timing model for execution-time analysis. *International Workshop on Real-Time Embedded Systems RTES 2001 (held with 22nd IEEE RTSS 2001), London, UK.*

Au, P., Darlington, J., Ghanem, M., Guo, Y., To, H. W., and Yang, J. (1996). Co-ordinating heterogeneous parallel computation. In *Euro-Par, Vol. I*, pages 601–614.

Bate, I., Bernat, G., Murphy, G., and Puschner, P. (2000). Low-level analysis of a portable wcet analysis framework. In *6th IEEE Real-Time Computing Systems and Applications (RTCSA2000)*, pages 39–48.

Berman, F. et al. (2001). The GrADS project: Software support for high-level Grid application development. *Int. J. of High Performance Computing Applications*, 15(4):327–344.

Bernat, G., Burns, A., and Wellings, A. (2000). Portable worst case execution time analysis using java byte code. In *Proc. 12th EUROMICRO conference on Real-time Systems*.

Bischof, H. and Gorlatch, S. (2001). Parallelization tridiagonal system solver by adjustment to a homomorphic skeleton. In Jähnichen, S. and Zhou, X., editors, *Proceedings of the Fourth International Workshop on Advanced Parallel Processing Technologies*, pages 9–18.

Foster, I. and Kesselmann, C., editors (1998). *The Grid: Blueprint for a New Computing Infrastructure*. Morgan Kaufmann.

Kennedy, K. et al. (2002). Toward a framework for preparing and executing adaptive grid programs. In *Proceedings of NSF Next Generation Systems Program Workshop (International Parallel and Distributed Processing Symposium 2002)*, Fort Lauderdale.

Meyer, J. and Downing, T. (1997). *Java Virtual Machine*. O'Reilly.

Press, W. H., Teukolsky, S. A., Vetterling, W. T., and Flannery, B. P. (1992). *Numerical Recipes in C: The Art of Scientific Computing*. Cambridge University Press, second edition.

A COORDINATION MODEL TO ESTABLISH ELECTRONIC COMMERCE POLICIES

Shundong Xia, Ruonan Rao, and Jinyuan You
Department of Computer Science
Shanghai Jiaotong University
Shanghai 200030, P.R.China
sdxia922@sina.com, rao-ruonan@cs.sjtu.edu.cn, you-jy@cs.sjtu.edu.cn

Abstract   Modern Electronic Commerce environment are heavily based on internet and involve issues such as distributed execution, multiuser interactive access or interface with and use of middleware platforms. Thus, their components exhibit the properties of communication, cooperation and coordination as in CSCW, groupware or workflow management systems. In this paper, we examine the potential of using SIKA, a coordination model developed by us, to establish electronic commerce policies and we show the benefits of such an approach.

Keywords: Coordination Models, Coordination Languages, Electronic Commerce, Tuple Spaces

1.   Introduction

Research on electronic-commerce focused mainly on means for secure and efficient transfer of funds [7, 2, 4]. Such means are indeed necessary for e-commerce, but they are not sufficient.

Commercial activities are not limited to simple exchange of funds and merchandise between a client and a vendor - they often consist of multistep transactions, sometimes involving several participants, that need to be carried out in accordance to a certain policy. A commercial policy is the embodiment of a contract (which may be explicit or implicit) between the principals involved in a certain type of commercial activity.

The currently prevailing method for establishing e-commerce policies is to build an interface that implements a desired policy, and distribute this interface among all who may need to operate under it. Unfortunately, such "manual" implementation of e-commerce policies is both unwieldy and unsafe. It is unwieldy in that it is time consuming and

G. Hommel and S. Huanye (eds.), The Internet Challenge: Technology and Applications, 121–128.
© 2002 *Kluwer Academic Publishers.*

122

expensive to carry out, and because the policy being implemented by a given set of interfaces is obscure, being embedded into the code of the interface. A manually implemented policy is unsafe because it can be circumvented by any participant in a given commercial transaction, by modifying his interface for the policy.

Coordination has been defined as the process of managing dependencies between activities [3]. In the field of Computer Science coordination is often defined as the process of separating computation from communication concerns [5]. SIKA (Secure InterlinK Architecture) is a coordination model we developed recently to provide a coordination infrastructure for network computing environment [8]. In this paper, we show that SIKA is a preferable mechanism to establish e-commerce policies.

2.   The Architecture of SIKA

2.1   Modules in SIKA

The whole structure of SIKA can been seen in *Figure 1*. The dispatcher works as an interface between SIKA and agents. The group manager is in charge of creating and revoking coordination groups, managing the memberships of agents. A coordination group links a set of autonomous agents which have a coordination relationship. The administration console provides an interface to administrator to control the group manager and the dispatcher.

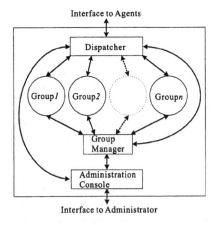

Figure 1.   Modules in SIKA

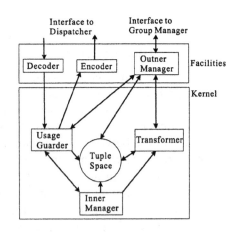

Figure 2.   Architecture of Coordination Group

2.2   Cryptosystem

All the cryptology used in this paper is based on RSA public-key system [6]. $E(x, \{m\})$ represents encrypting m using the key x. The administrator has a public key Y_c and a private key X_c. An agent A has a public key Y_A and private key X_A.

2.3   Coordination

The architecture of a coordination group can be seen in *Figure 2*. When in coordination, the agent A can send the message $(I_A, E(Y_c, \{E(X_A, \{operation\ primitives\})\}))$ to the dispatcher which will forward it to the corresponding group. Operation primitives used in SIKA include:

1) `create_task(task type)`: Create a task.

2) `in(task,template)`: In the "task", asynchronously retrieve a tuple which matches the "template" and remove the tuple from the tuple space.

3) `rd(task,template)`: This is the same as "in" except that the tuple is not removed.

4) `in_b(task,template)`: In the "task", synchronously retrieve a tuple which matches the "template" and remove the tuple from the tuple space.

5) `rd_b(task,template)`: This is the same as "in_b" except that tuple is not removed.

6) `out(task,tuple)`: In the "task", insert the "tuple" into the tuple space.

The usage guarder checks if the operation is legal according to the coordination policy. Any result R will be sent to the encoder. Then the encoder send the message $(I_A, E(Y_A, \{E(X_c, \{R\})\}))$ to the dispatcher. The dispatcher will return the message to corresponding agent.

The transformer will transform corresponding tuples according to the coordination policy.

The inner manager is used to capture the triggered events and change the state of the group.

3.   SIKA Coordination Language (SIKA_L)

SIKA coordination language (SIKA_L) is used to specify coordination policy. There are four parts in the coordination policy specified by SIKA_L: objects declaration, usage guarder, transformer and manager.

3.1   Objects declaration

The syntax of roles declaration in SIKA_L can be seen in *Figure 3*.

```
<roles declaration>::="role" {<role name>["["<count>"]"]["":"<max>]","}*
                       <role name>["["<cout>"]"]["":"<max>]";"
<tuples/templates declaration>::=
         "data"{<name>["("{<type><name>","}*<type><name>")"]","}*
               <name>["("{<type><name>","}*<type><name>")"]";"
<events declaration>::=
         "event"{<name>["("{<type><name>","}*<type><name>")"]","}*
               <name>["("{<type><name>","}*<type><name>")"]";"
```

Figure 3.   Syntax of Objects declaration

A Role is an abstract representation of a same class of agents. <count> is used to declare a number of homogeneous roles. <max> represents the maximal number of agents which can play the role.

3.2   Usage guarder

The usage guarder works according to the coordination policy specified by the syntax in *Figure 4*.

```
<usage guarder>::="guarder{"[<variable declaration>]{<usage>}*"}"
<usage>::=<role name>["["<variable>"]"]"{"{<task>";"}*"}"
<task>::=<task name>["["<count1>"]"]["{disable}"]":"<operation expression>
<operation expression>::=
       <operation primitive>["{pre:"<event>"}"]["{post:"<event>"}"]
       |<operation expression>"."<operation expression>
       |<operation expression>"|"<operation expression>
       |"{"<operation expression>"}"[<count2>|"*"]
<operation primitive>::=<operator>"("<template>")"
<operator>::="in"|"rd"|"in_b"|"rd_b"|"out"
```

Figure 4.   Syntax of usage guarder

An agent can only create at most <count1> homogeneous tasks.

An <operation expression> specifies the legal sequence of operations.

<operation expression>.<operation expression> means that the corresponding agents must orderly execute the two sequence of operation primitives.

<operation expression>|<operation expression> means that the corresponding agents can only choose to execute one of the two sequences of operation primitives.

{`<operation expression>`}[`<count2>`|`*`] means that when an agent has done the sequence of operation primitives in the expression successfully, it can do it again. `<count2>` specifies the repeat number. "`*`" means that the repeat number is infinite.

{`pre:<event>`} specifies the event trigged before the agent do the legal operation primitive. Likewise, {`post:<event>`} specifies the event trigged after the agent do the legal operation primitive.

3.3   Transformer

The transformer works according to the coordination policy specified by the syntax in *Figure 5*.

```
<transformer>::="transformer{"[<variable declaration>]{<rule>";"}*"}"
<rule>::=<rule name>[<variable list>]":"
      <left>"->"<right>["{"<condition>"}"]
<left>::={<template>","}*<template>
<right>::={<template>","}*<template>
```

Figure 5.   Syntax of transformer

A transformer contains several rules. Several instances can be created by the inner manager from a rule. An active instance in which the condition is satisfied can transform the tuple space by replacing the tuples match the templates in left part with the tuples generated by the templates in the right part.

3.4   Manager

the inner manager works according to the coordination policy specified by the syntax in *Figure 6*.

Besides the ordinary events generated by agents when they execute legal operation primitives, the inner manager also catches build-in events. "initiation", which is one of the build-in events, generated when the coordination group is created.

Once the inner manager catches an event some reactions which will change the state of the coordination group.

```
<Manager>::="manager{"[<variable declaration>]
        {<extended event>":{"{<reaction>";"}*"}"}*"}"
<extended event>::=<event>|<build-in event>
<build-in event>::=
        "initiation"
        |<template>"."{"remove"|"add"}{"{pre}"|"{post}"}
        |<role name>["["<variable name>"]"]"."{"join"|"leave"}{"{pre}"|"{post}"}
<reaction>::=
        {"enable("|"disable("|"restart("|"guarder."
                <role name>"."<task name>["["<variable name>"]"]")"
        |{"enable("|"disable("}"transformer."
                <rule name>"."["("<variable list>")"]")"
        |{"add("|"remove("}<template>")"
        |{"add("|"remove("}"transformer."<rule name>["("<variable list>")"]
        |"set("<variable name>","<expression>")"
        |<condition>"{"{<reaction>";"}*"}"
        |"("<variable name>"="<initial value>".."<final value>")"
                "{"{<reaction>";"}*"}"
```

Figure 6.   Syntax of manager

4.   Establish Electronic Commerce Policies Using SIKA

Now we'll give an example for establishing electronic commerce policy. Consider a distributed database that charges a fixed and small fee per query. A prospective client c can purchase from a designated ticket-seller a non-copyable q-valued ticket, where q is the number of queries this ticket is good for. A query submitted by c is processed only if c has a ticket with non-zero value, and as a result the value of this ticket is reduced by one. The policy specified by SIKA_L can be seen in *Figure 7*.

In *Figure 7*, _selfID(), _selftask(),_creator(tuple_r t), _task(tuple_r t) and _tuple(int i) are all build-in functions. _selfID() returns the id of the agent in current context. _selftask() returns the id of the task in current context. _creator(tuple_r t) returns the id of the agent which create the tuple referenced by t. _task(tuple_r t) returns the id of the task which create the tuple referenced by t. _tuple(int i) returns the reference to the i-th tuple of the left part in current transformer rule.

First, a client creates a "buy" task and adds a tuple "buyticket(quantity, EC)" to the tuple space (line 13). Here "EC" is a certificate for a sum of dollars in some form of electronic cash [1]. The transformer transforms the tuple by adding the client's id (line 28). The ticket-seller processes the tuple (line 19) and returns a tuple "giveticket(agent_id, quantity)" (line 20). Then the client can read the tuple which contains

```
1     role client:100,ticketseller;
2     data buyticket(int quantity,string EC),
3          buyticket2(string agent_id,int quantity,string EC),
4          giveticket(string agent_id,int quantity),
5          ticket(string agent_id,int quantity),
6          query(string statement),
7          query2(string agent_id,string task_id,string statement),
8          answer(string agent_id,string task_id,string result);
9     event ack(string agent_id);
10    guarder{
11      string agent_id,task_id;
12      client{
13         buy:{out(buyticket(_,_)).
14               rd_b(giveticket(_selfID(),_)){post:ack(_selfID())}}*;
15         query[10]:{out(query(_)).
16                    in_b(answer(_selfID(),_selftask(),_))}*;
17      }
18      ticketseller{
19         sell:{in_b(buyticket2(agent_id,_,_)).
20              out(giveticket(agent_id,_))}*;
21         serve:{in_b(query2(agent_id,task_id,_)).
22              out(answer(agent_id,task_id,_))}*;
23      }
24    }
25    transformer{
26      string agent_id,EC,statement;
27      int N1,N2;
28      r1:buyticket(N1,EC)->buyticket2(_creator(_tuple(1)),N1,EC);
29      r2(agent_id):giveticket(agent_id,N1)->
30                   ticket(agent_id,N1);
31      r3:ticket(agent_id,N1),ticket(agent_id,N2)->
32                   ticket(agent_id,N1+N2);
33      r4:query(statement),ticket(_creator(_tuple(1)),N1)->
34           query2(_creator(_tuple(1)),_task(_tuple(1)),statement),
35           ticket(_creator(_tuple(1)),N1-1);
36      r5:ticket(agent_id,0)->;
37    }
38    manager{
39      string agent_id;
40      initiation:{add(transformer.r1);enable(transformer.r1);
41                  add(transformer.r3);enable(transformer.r3);
42                  add(transformer.r4);enable(transformer.r4);
43                  add(transformer.r5);enable(transformer.r5);}
44      ack(agent_id):{add(transformer. r2(agent_id))
45                  enable(transformer.r2(agent_id)),
46                  remove(transformer.r2(agent_id));}
47    }
```

Figure 7.   Ticket policy

the id of the client (line 14). Immediately after the agent read the tuple, an event "ack(agent_id)" is triggered. The inner manger catches the event and activates transformation rule "r2"(line 44-46). Rule "r2" transforms the tuple to a "ticket" (line 29-30). If there has existed an old "ticket" belonged to the same agent, the old one and the new one will combine (line 31-32).

After a client adds a tuple "query(statement)" to the tuple space (line 15), the transformer determines if the client has a ticket (line 33).

If the client has, the transformer generates a new tuple "query2" which is available to the ticket-seller (line 34), and reduces the value of the ticket by one (line 35). Immediately after the value of a ticket becomes zero, the ticket is deleted (line 36). The ticket-seller processes the tuple "query2" (line 21) and returns the result (line 22). At last, the client takes the result belongs to it (line 16).

5. Conclusion

We have introduced SIKA and its application to electronic commerce. An electronic commerce policy established by SIKA enjoys a number of desirable properties.

1) Asynchronous. Participants coordinated by SIKA need not be temporally coupled.

2) Anonymous. Participants need not know the network addresses of others.

3) Secure. The communications between participants and SIKA are protected by cryptosystem. The behavior of each participants is constrained by the usage guarder.

4) Flexible. The e-commerce policy can be easily changed in design time and run time.

References

[1] Fan C.I., Chen W.K., and Yeh Y.S., "Date attachable electronic cash", *Computer Communications* 23, no. 4 (2000) 425–428.

[2] Glassman S., Manasse M., Abadi M., Gauthier P., and Sobalvarro P., "The Millicent protocol for inexpensive electronic commerce", *Proc. 4th International World Wide Web Conference*, Boston, USA, O' Reilly & Associates, Inc., 1995, 603–618.

[3] Malone T. W. and Crowston K., "The interdisciplinary study of coordination", *ACM Computing Surveys* 26, no. 1 (1994) 87–119.

[4] Panurach P., "Money in electronic commerce: Digital cash, electronic fund transfer and ecash", *Communications of the ACM* 39, no. 6 (1996) 45–50.

[5] Papadopoulos G. A. and Arbab F., "Coordination models and languages", in *Advances in Computers*, Zelkowitz M. V. Editor, Academic Press, San Diego, 1998, 329–400.

[6] Rivest R., Shamir A., and Adleman L., "A method for obtaining digital signatures and public key cryptosystems", *Communications of the ACM* 21, no. 2 (1978) 120–126.

[7] Sirbu M. and Tygar J.D., "Netbill: An Internet commerce system", *Proc. IEEE COMPCON*, San Francisco, USA, IEEE Press, 1995, 20–25.

[8] Singh M.P., Vouk M.A., "Network computing", in *Encyclopedia of Electrical and Electronics Engineering*, Webster J.G. Editor, John Wiley & Sons, Inc., New York, 1999, 114–132.

TOWARDS AN INTERNET-BASED MARKETPLACE FOR SOFTWARE COMPONENTS

Kurt Geihs
TU Berlin, Germany [geihs@ivs.tu-berlin.de]

Abstract: Software components provide support for reusability, extensibility, and flexibility which are required in distributed application scenarios. Lately, a new kind of software component has attracted considerable attention, i.e. web services. A web service is a self-contained, self-describing modular application component that can be published, located and invoked across the web. We claim that this trend will lead to a "component marketplace" where software components are traded. Depending on the type of component and the application requirements the spectrum of component usage will range from purchasing and downloading a component as a whole to leasing and accessing a component over the Web. A market infrastructure is needed in order to dynamically offer, trade, bind and deploy software components.

Key words: software component, web service, electronic market, middleware, trading

1. INTRODUCTION

Software components provide support for reusability, extensibility, and flexibility which are required in distributed application scenarios. The goal is to compose applications out of software building blocks similar to what we are used to when assembling computers from standardized hardware components. Lately, a new kind of software components has attracted widespread attention, i.e. web services. A web service is commonly defined to be a self-contained, self-describing modular application component that can be published, located and invoked across the Web.

G. Hommel and S. Huanye (eds.), The Internet Challenge: Technology and Applications, 129–138.
© 2002 *Kluwer Academic Publishers.*

Although the web service technology is still in its infancy, it has gained a lot of attention already. Practically all vendors emphasize the importance of web services for application integration tasks and for service oriented architectures. Various development tools for web services are being integrated into software development environments. They ease the development of web service components considerably by making much of the interface specification and encoding transparent.

Looking at the trends towards component-based software and web services (as components over the web) the situation arises that application developers have a choice between importing a component as a piece of software and leasing a component over the web. Clearly, this choice depends on the application context, but it is not hard to envisage services that may either be provided locally or remotely, e.g. information services, language dictionaries and financial calculations. The discussion on application outsourcing and application hosting points to this direction, too. With web services as components over the web the granularity of outsourcing is decreased.

When components are tradable goods that may be purchased or leased we will see markets for components emerge where component providers and users (i.e. software developers) meet and negotiate. To make such a vision come true we need a new technical market infrastructure that is tailored to the particular requirements of software components. In this paper we develop this challenging vision and elaborate on the required ingredients of an appropriate market infrastructure. Thus, the paper is a discussion paper; it does not yet present a comprehensive solution.

The paper is organized as follows. Section 2 highlights the basics of component-based software systems. In Section 3 we present briefly some web service fundamentals. Section 4 analyses the requirements of a market infrastructure for components and discusses the major building blocks as well as their design alternatives. Section 5 contains our conclusions and outlook.

2.   COMPONENT-BASED SOFTWARE

Traditional mainframe applications typically were rather closed systems that were focused on a single application function, e.g. order processing or inventory management. Today, enterprise applications are viewed mostly as a chain of integrated activities, where individual software modules need to interoperate in a coordinated fashion. Obviously, such application systems are inherently distributed. Scalability is one of the most important requirements. Applications should be extensible and adaptable to their ever changing environment. The need for integration does not stop at the borders of an enterprise. Inter-enterprise application integration manifests itself in collaborative business-to-business applications, such as supply-chain management systems and e-business marketplaces.

In order to satisfy the required interoperability and integration of applications we need to break up the monolithic application architectures of the past and replace them with multi-tier architectures that ease the distribution and support modularity and flexibility. Object-oriented software design and programming was a major step in software technology to respond to these needs. The object model has become a widely accepted foundation for distributed application systems.

However, the object-oriented programming paradigm alone with its rather fine-grained view of the world and its strong emphasis on the invocation interface of an object has not solved all the problems of large scale application development for distributed environments. Reusability and composition require more than such a fine-granular programming model. In order to reuse a piece of software we need to understand (and specify) its dependencies on other modules and on its operating environment, its error behavior, how it is created and deleted, and much more. Software component models and their associated frameworks try to manage this information explicitly.

A software component is "a unit of composition with contractually specified interfaces and explicit context dependencies only. A software component can be deployed independently and is subject to composition by third parties" [Szyp97]. Popular examples of component frameworks for enterprise-level applications are Enterprise Java Beans [Java], the CORBA Component Model (CCM) [CORBA3] and .NET by Microsoft [.NET]. EJBs are language-dependent (i.e. on Java) but platform independent. The .NET framework supports many different programming languages, but is heavily focused on Microsoft Windows platforms, although core elements of .NET have been ported to other environments, too. (There is a Free-BSD port publicly available.) The proposed CCM, which is compatible with the EJB framework, claims to be language- and platform-independent, but has yet to demonstrate whether this can be achieved. Furthermore, the commercial relevance of the CCM is rather unclear, since – as far as we now – no vendor is developing an implementation currently.

There are a number of research activities that investigate aspects of dependability and quality in the design and implementation of software components. Just like hardware components come with a technical specification, which describes in detail the properties and operating conditions, a software developer will want to know the properties of a software component that she wants to reuse. For example, in the QCCS

132

project we address these non-functional aspects by means of contractual guarantees for component properties and aspect-oriented programming techniques [QCCS].

3. COMPONENTS OVER THE WEB

Lately, a new kind of component has been introduced: web services in the Internet. A set of standards has been developed that define a web service as a self-contained, self-describing modular application component that can be published, located and invoked across the Web [C02].

Invocation of a web component is done using SOAP, the Simple Object Access Protocol, which is build on XML for data description and HTTP for invocation transport (other standardized protocols are possible). Service interfaces and properties are described in WSDL (Web Service Description Language), which is also an XML application. UDDI (Universal Description, Discovery and Integration) provides a rudimentary directory service, where service providers can register their services and where customers can find out about service offers. Figure 1 illustrates these inter-actions: In Step 1 two servers register their service offers at an UDDI directory, in Step 2 the client inquires about suitable service providers and receives an URL that points to the service description of Server2, in Step 3 the client retrieves the service description of that server, and finally in Step 4 the client invokes the web service. All interactions are carried out using SOAP as the invocation protocol.

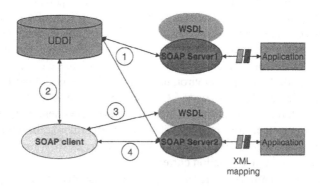

Figure -1. Web Service Interactions

In terms of the CORBA terminology, WSDL corresponds to the CORBA IDL, SOAP to the CORBA IIOP, and UDDI can be seen as a special kind of naming and trading system.

Web services may provide a quite diverse spectrum of service types, from business oriented (e.g. credit authorization), over consumer oriented (e.g. stock quotes) to more system oriented (e.g. user authentication). The web service technology seems to encounter broad acceptance. All major component frameworks support the integration of web services into their component-based applications.

While SOAP, WSDL and UDDI are first steps to make the Internet a special kind of middleware platform, it is obvious that – just like in middleware systems such as CORBA and COM+ – distributed infrastructure services such as security and transactions have to be integrated into the web service technology. Some standardization activities have been started, but results are not yet available. Furthermore, as is the case with regular software components – or perhaps even more so – the developer of an application will want to know about the quality attributes of a web service, e.g. its availability or response times guarantees. We need technical means to negotiate such "service level agreements" and to monitor the service provision during runtime in a trustworthy way.

There are various research, development and standardization activities going on that address these concerns [OASIS].

4. COMPONENT MARKET

In a marketplace for components we need to distinguish three actors, i.e. the customer, the provider, and the common distribution infrastructure. Our emphasis in the following discussions is on the common infrastructure that supports the market activities. Key design aspects of such an infrastructure are:

- techniques for the specification of configuration and context dependencies as well as quality requirements of components;
- trading support and repositories for component offers;
- negotiation of usage contracts between provider and customer;
- monitoring component usage at run-time;
- support for accounting and billing according to usage contracts;
- security.

4.1 Specification

In order to specify desired non-functional properties of components we need means to separate the specification of functional aspects (i.e. the business logic) from non-functional ones, such as reponse-time, fault-tolerance and so on. Commonly, this is called "separation of concerns" in the realm of software engineering. The Unified Modeling Language (UML) appears to be a suitable candidate for such a specification language, since it is the design notation of choice for many software developers. It gives the component designer a rich, but somehow disorganized, set of views on her model as well as many features, such as design pattern occurrences, stereotypes or tag values to add non-functional annotations to a model.

Thus, it takes a well coordinated approach when applying UML as the component specification language. We need to organize the existing UML features around the central notions of quality of service contracts for specifying non-functional properties and the support for their implementation. In the project QCCS we investigate such an UML-based approach to quality-controlled components [QCCS].

4.2   Trading

Conventional trading services [ISO, OMG] have adopted a centralized trading model: The Trader acts as a central matchmaker for customers and providers. Primarily, it offers at its interface two operations, i.e. export() and import(). A service provider advertises its service offer by calling the export operation on the trader. The export operation includes at least the specification of the service type and a service interface identifier which is used by clients to access the service provider. Optionally, it may contain additional service attribute values. The trader maintains a service repository in which it stores the available service exports. A service requester calls the import operation on the trader specifying the desired service type and optionally some service attribute values. The trader may interact with other traders to engage in a trading federation and thus enlarge its repository.

Opposed to this centralized, strict client/server model, peer-to-peer and publish/subscribe techniques have proven to be effective means of information dissemination. This should be exploited in a component market, too. Component users and providers form a peer-to-peer network where information on available components is forwarded according to some well-known peer-to-peer protocol, such as Gnutella [Oram01] or Freenet [CSWH01]. Such a dissemination strategy corresponds much better to the objective of component reuse, because every user becomes a potential provider to others, very similar to the exchange of music files in current peer-to-peer systems.

Furthermore, we need to support a somewhat fuzzy matching scheme in the search for components. Partial matches of request and offer may be acceptable to the component requester in certain cases. Thus, a technique for evaluating and quantifying the degree of overlap between XML service descriptions is needed. Here one can build on existing research results (e.g. [TW00]).

4.3   Negotiation

If a customer has found an appropriate component provider, there will be a negotiation between the two about the type of usage (buy or lease) and the usage conditions (quality of service, duration, price, etc.). A multi-stage negotiation with several negotiation rounds may be required. Finally, the negotiations will lead to some kind of contract between customer and provider. This contract must be monitored during the usage of the component. In complex application scenarios involving many components, it may be advisable for customers to aim for compound contracts with several components in one package.

We have studied similar negotiations in the context of quality of service management in middleware [BGG99].

4.4   Monitoring

During web component usage the agreed QoS level has to be monitored to detect violations. Quality of service is an end-to-end issue. It is perceived and judged primarily by the client. Consequently, the monitoring should be done at the client side. Depending on the quality characteristics, part of the monitoring activity can be delegated to the distribution infrastructure that connects client and server. For example, response time must be measured from the client perspective, but the infrastructure could measure the influence of transmission errors and retransmissions on the overall response time.

Upon violation of a QoS agreement the client application can either adapt its behavior to the new situation or renegotiate. The information about the duration of a certain QoS agreement may be input for the accounting service.

4.5   Accounting

In addition to one-time purchases, components over the web provide software vendors with the option of selling their application components to clients over the Internet on a per-use or subscription basis. A reliable, yet easy-to-use and scalable accounting system for component usage will be a major prerequisite to make a component marketplace reality. A logically centralized accounting agency could be part of the market infrastructure. It receives and collects component usage data from the monitoring mechanisms. The usage data could be transformed into bills to customers for their component usage. A major difficulty will be to agree on a widely accepted electronic payment scheme.

Since components may be bought or leased, customers need to carefully evaluate their usage requirements and the related cost implications. Likewise, component providers need to find competitive cost models for component usage.

4.6   Security

In a component marketplace we need to distinguish of web service components that are used temporarily and the security of components that are purchased and integrated locally as a whole. We assume that in both cases the user/developer knows and understands the semantics and side effects of the component. This is not different from using a remote service or reusing some software module in today's IT environments.

The security requirements of web services are basically the same as for any other service in distributed systems. For example, the CORBA Security Service specification [OMG98] contains a detailed analysis of security threats in distributed systems and possible remedies. Web services pose one additional problem: typically, a web service is contacted via TCP-port 80, which is the well-known port of web servers. Therefore web service traffic passes freely through corporate firewalls. It takes a special kind of "web service firewall" in order to protect a system against malicious attacks. There are activities going on that try to develop a specific security framework for web services [OASIS]. They focus on the definition and exchange of authentication and authorization information which is encoded in a platform-independent way. A rather novel approach is called "content-based access control" where security decisions are based on the contents of a message.

If components are acquired on a component market and integrated locally into applications, the security requirements are quite different. To a certain degree, they resemble the security concerns in mobile agent platforms [ZMG98]. The user needs to establish trust into the author and sender of a component. Therefore, a component must have some form of digital signature that clearly identifies its origin. Different levels of trust may be in place such that component sources are classified according to their trustworthiness and reputation.

5.   CONCLUSIONS

Component-based software engineering including "components over the web" opens up challenging and unique research problems. We have presented our vision of an emerging software component marketplace where components may be purchased or leased over the Internet. Although many technological pieces of the required marketplace infrastructure are available today, a coherent, integrated approach is lacking and more innovations are needed to make this vision a reality.

With such a marketplace for components the nature of software development will change. The integration of reusable components poses new requirements to the specification and construction of web-oriented applications. To ensure high quality for these applications we need novel techniques to engineer and maintain desired non-functional properties such as reliability, scalability, security, and performance.

ACKNOWLEDGEMENTS

The author thanks H. König and H. Krumm for fruitful discussions.

REFERENCES

[.NET]   Microsoft .NET, http://www.microsoft.net/net
[BGG99]   C. Becker, K. Geihs, J. Gramberg, "Representing Quality of Service Preferences by Hierarchies of Contracts", in: Steiner, Dittmar, Willinsky (eds.) „Elektronische Dienstleistungswirtschaft und Financial Engineering", Schüling Verlag, Münster (1999)
[C02]   F. Curbera et.al., "Unraveling the Web Services Web", IEEE Internet Computing, pp. 86-93, March/April 2002
[CORBA3]   J. Siegel, "An Overview of CORBA 3", Proc. 2nd IFIP Int'l Working Conf. Distributed Applications and Interoperable Systems (DAIS 99), Helsinki (1999)
[CSWH01]   I. Clarke, O. Sandberg, B. Wiley, and T. Hong, "Freenet: A distributed anonymous information storage and retrieval system", in: Designing Privacy Enhancing Technologies: International Workshop on Design Issues in Anonymity and Unobservability (New York, 2001), H. Federrath, Ed., Springer.
[ISO]   International Standardization Organization (ISO): ODP Trading Function, IS 13235
[Java]   http://java.sun.com
[OASIS]   http://www.oasis-open.org
[OMG96]   OMG Trading Object Service, OMG Document orbos/96-05-06
[OMG98]   OMG Security Service, Rev. 1.2, Object Management Group (1998)
[Oram01]   A. Oram (Ed.), "Peer-to-Peer, Harnessing the Power of Disruptive Technologies", O'Reilly Books, 2001.

[QCCS]   T. Weis, C. Becker, K. Geihs und N. Plouzeau, "An UML Meta-Model for Contract aware Components", Proc. of UML 2001, Toronto/Canada (2001)

[Szyp97] C. Szyperski, "Component Software - Beyond Object-Oriented Programming", Addison-Wesley (1997)

[TW00]   A. Theobald, G. Weikum, "Adding Relevance to XML", Proceedings of the 3rd International Workshop on the Web and Databases, LNCS, Springer (2000)

[ZMG98]   M. Zapf, H. Müller, K. Geihs, "Security Requirements for Mobile Agents in Electronic Markets", Proc. of IFIP Working Conference on Trends in Distributed Systems for Electronic Commerce (TrEC'98), LNCS 1402, Springer (1998)

M-SCTP: DESIGN AND PROTOTYPICAL IMPLEMENTATION OF AN SCTP-BASED, END-TO-END MOBILITY CONCEPT FOR IP NETWORKS

Wei Xing, Holger Karl, Adam Wolisz
Technical University Berlin, Telecommunication Networks Group,
Einsteinufer 25, 10587 Berlin, Germany
xing|karl|wolisz@ee.tu-berlin.de

Harald Müller
Siemens AG, Information and Communication Networks
München, Germany

Abstract   The problem of mobility in IP networks has traditionally been solved at the network layer. We present an alternative solution that solves it at the transport layer in an end-to-end fashion, leveraging the ability of a modern transport layer protocol (SCTP) to use multiple IP addresses per association. This is achieved by dynamically modifying this set of IP addresses with new IP addresses that are assigned to the mobile node as it moves around the network We show that this is a feasible approach, discuss the performance ramifications of underlying protocol mechanisms and identify those that are deterrent to handoff performance.

Keywords: SCTP, mobility, IP, transport-layer, handoff, performance

1.   Introduction

The proliferation of laptops, hand-held computers, cellular phones, and other mobile computing platforms connected to the Internet has triggered much research on mobility support in IP networks. Modern mobile terminals are likely to be equipped with wireless communication devices allowing them to constantly reach the Internet and participate in it as normal end systems would. In the near future, a mobile terminal will be able to select among several concurrently available, different wireless technologies the one which best supports the

G. Hommel and S. Huanye (eds.), The Internet Challenge: Technology and Applications, 139–147.
© 2002 *Kluwer Academic Publishers.*

user's current communication needs and offers the best cost tradeoffs or even to use multiple access points to the fixed network infrastructure in parallel.

Such a mobile terminal, however, represents an ill fit with the traditional assumptions upon which the IP protocols is based: a classic end system does not move and has only a single point of attachment to the network. For such an end system, a single handle is sufficient to represent both the identity of a terminal as well as its location within the network; the IP address is this very handle. Such a permanent handle is not appropriate in mobile networks. On one hand, an identifier is necessary to distinguish among different terminals, on the other hand, information about the current location within a network has to be provided to ensure that packets destined to a certain terminal can still be routed towards this terminal. The fundamental mobility problem in IP-based networks is therefore the separation of identity and location.

Current approaches, e.g., Mobile IP [2], solve this problem at the network layer. While general, these approaches have some unappealing characteristics; limited performance and additional complexity for the network architecture are perhaps the two most serious shortcomings. Adding complexity to the network runs counter to some of the basic design principles of the Internet, in particular, the end-to-end principle [4]: anything that can be done in the end system should be done there. And in fact, supporting mobility is, strictly speaking, an end system issue. The question hence arises whether it would be possible and beneficial to attempt to support mobility in IP networks at higher layers than the network layer. The lowest end-to-end layer, in the Internet protocol stack, is the transport layer. As the transport layer is considerably affected by mobility—e.g., it has to be able to quickly adapt its flow and congestion control parameters to the new network situations during and after handovers—it is a natural candidate for mobility support. Attempting to support mobility in the transport layer might enable it to improve its parameter adaptation, it could also make the entire networking architecture simpler by working without additional entities within the network.

Implementing such a concept requires changes to existing transport layers. In previous work, TCP has been modified to support such an end-to-end mobility concept. While TCP is indeed the most often used transport protocol in the Internet, it might not be the perfect platform to experiment with unconventional ways of supporting mobility. In particular, when considering the potential that a mobile terminal could be in contact with multiple access points at the same time, other protocols might offer a simpler starting point. A good candidate is the Stream Control Transmission Protocol (SCTP): an SCTP "association" (essentially, a connection) can use multiple addresses simultaneously. While this property was not originally intended to support mobility (the rationale is to support highly available servers), it presents an excellent platform on which to experiment with new mobility-support mechanisms. In addition, many of its basic mechanisms such as flow and congestion control are very similar to TCP. Therefore, we decided to use SCTP as a starting point and introduce mobility support for it. This paper presents our architecture as well as a prototype implementation along with preliminary performance results.

The remainder of this paper is structured as follows: Section 2 gives an overview of the most relevant related work. Section 3 introduces necessary SCTP basics, Section 4 describes our proposed architecture to enable SCTP to support mobile terminals. A prototypical implementation of this architecture is introduced in Section 5 along with some preliminary performance evaluation results. Finally, Section 6 contains our conclusions and options for future work.

2. Related Work

Several architectures have been proposed to provide IP mobility support; Mobile IP (MIP) [2] is probably most widely known. Mobile IP uses a couple of addresses to manage user's movements. Each time the mobile host (MH) connects to the so-called foreign network, it obtains a temporary address called Care-of-Address (CoA) from a mobile agent in the local network called the foreign agent (FA). The MH must inform its home agent (HA) of this new address by the registration process. The home agent then assumes the MH's permanent IP address. Once a packet destined to the MH arrives at the home agent, it is tunneled towards the MH using the MH's temporary CoA.

The advantage of MIP is a fully-transparent mobility support, which is general and sufficient for many mobile applications. But MIP also has some disadvantages. The main problems are handover latency and opaqueness for the transport layer. While this is in line with general information hiding principles, it also limits optimization potential.

An end-to-end architecture using TCP [5] is proposed for IP mobility, based on dynamic DNS updates. Whenever the mobile host moves to a new network, it obtains a new IP address and updates the DNS mapping for its host name—the separation of identity and location is here achieved using names and addresses, not two different addresses as it is done in Mobile IP. Merely changing the mobile terminal's address, however, would disrupt any ongoing TCP connection whenever a handover occurs. To overcome this problem, a new option, called Migrate TCP, was added and it is used to "migrate" an existing connection from the mobile terminal's old to the new IP address. In a TCP connection migration process, the MH activates a previously-established TCP connection from a new address by sending a special Migrate SYN packet that contains a token identifying this very connection. The correspondent host (CH) will then re-synchronize the connection with the MN using the new address. The time of a TCP connection migration process is called TCP migration delay in this paper. The main advantage of this approach is that it has no need for a third party (home agent) for smooth handoff. However, we believe that such architecture incurs similar handoff delays to those experienced in MIP, or even worse due to DNS update delays and migration delays. The main shortcoming of this TCP extension is its inability to use concurrently available access points: it is this shortcoming that we are especially interested in in overcoming. A similar approach has been proposed in [3], yet this work does not present a prototype implementation or performance results.

In conclusion, Table 1 shows that none of these approaches can fulfill all requirements: no third party, ability to use multiple access points and no modifications for intermediate routers. A mobility extension to SCTP is intended to satisfy all these three requirements.

	Third party requirement for handoff	Concurrent usage of access points	Modification for intermediate router
Mobile IP	Yes	No	Yes
Migrate TCP	No	No	No
SCTP mobility	No	Yes	No

Table 1. Comparison of different IP mobility approaches

3. SCTP Overview

Recently, a new transport layer protocol, "Stream Control Transmission Protocol" (SCTP) [6], has been proposed running on top of IP. It encompasses many basic functionalities of TCP, and adds a number of interesting protocol mechanisms. One core feature of SCTP is multi-homing, which enables a single SCTP endpoint to support multiple IP addresses *within a single association.*

This feature is a conceptually simple and powerful framework for IP mobility support at the transport layer as it already separates the identity of an end system from the current address to which packets are sent; it is a one-to-many correspondence as opposed to the one-to-one correspondence used by traditional transport layer protocols.

However, the multi-homing mechanism's purpose is to increase association reliability in wired networks. Therefore, the IP addresses of all involved end system are fixed and known in advance. SCTP can thus rely on all communicating peers to learn about all the IP addresses before the association is completely established, and these IP addresses must not be changed (either added or deleted) during the session in the classical SCTP. This does not work in a mobile environment as a mobile host does not have a fixed, previously known IP address: as it moves around, its local address constantly changes to reflect its new position in the IP routing hierarchy. To leverage SCTP's multi-homing mechanism, this set of IP addresses must be made dynamic.

4. The Mobile SCTP Architecture

The main idea of our approach is to let a mobile host have (at least) two IP addresses in the existing association during handoff if two access points are available simultaneously.

In order to provide IP mobility, the packets sent to a mobile host need to be forwarded to the new IP address in the new location visited by the MH without disrupting the current session. The basic idea is to exploit the case that the coverage areas of an old access point, which the mobile node currently uses, and the new access point, to which it will perform a handoff, are overlapping. Then, the mobile node can obtain an IP address from the new access point and use this new address to prepare the actual handoff process, i.e., to modify the set of IP addresses that describe a particular SCTP association in the correspondent host. This process is illustrated in Figure 1.

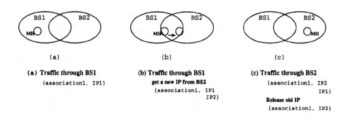

Figure 1. Mobile SCTP Mechanism

In more detail: A mobile host (MH) is in the coverage of BS1, as shown in Figure1 (a). In this case, the traffic is through BS1 and SCTP association1 has only one IP address (IP1), which it got from BS1. When the MH enters the area that is covered by both BS1 and BS2 (Figure1 (b)), it obtains a new IP address (IP2) from BS2, typically via Dynamic Host Configuration Protocol (DHCP) [1] or similar mechanisms. This new IP address will be immediately added into the current active association (association1). During this overlap time, it would even be possible to use both transmission paths (via both BS1 and BS2). After the MH enters the coverage of BS2 (Figure 1 (c)), traffic is through BS2 and here IP2 must be used in association1 because IP1 is no longer usable. A similar concept has been presented in [7].

An evident problem with such an approach is a movement of a mobile host that continuously results in its switching back and forth between both base stations. As long as the IP addresses from these base stations are still valid (i.e., their DHCP lease has not expired), it would be possible to reuse them and safe quite some signaling overhead. This requires the base stations to identify themselves to the mobile node so that the mobile node can recognize whether it is re-connecting to a base station that it has just been using; in addition, the mobile node needs to store base station identifiers and local IP addresses in order to be able to easily reuse these addresses.

5.   Mobile SCTP Prototype

5.1   Implementation

We have implemented M-SCTP under Linux based on a user-space SCTP implementation (available via http://www.sctp.de) Specifically, we implemented a module to support dynamically adding a new IP address into an existing SCTP association. An additional module is responsible for detecting the availability of new base stations; this module serves as a link-layer trigger mechanism for M-SCTP (cp. Figure 2). If a new base station is found, i.e., a new interface with a new IP address is available, a user-defined signal will be sent to the SCTP application. The ADD IP process will then be started to add the new IP address into current SCTP instance (Mobile Host). After the new IP address is successfully put into the SCTP instance, an ADD_IP chunk will be immediately generated and sent to the remote SCTP peer. When the SCTP peer get the ADD_IP chunk, an addip callback function will be triggered to start the addition of the peer's new IP address from the ADD_IP chunk into the current association. Once completed, an ADDIP-ACK chunk is generated, acknowledging the availability of the new IP address at the correspondent host. After the MH receives ADDIP-ACK chunk from the remote SCTP peer the add-ip process has finished; the new IP address can be used according to the SCTP path management mechanism which are responsible for detecting unavailable IP addresses and switching over to secondary IP addresses. In summary, this allows the handover process to be solved using existing multi-homing mechanisms.

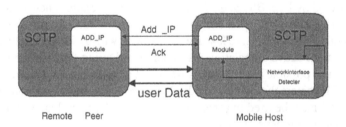

Figure 2.   M-SCTP Prototype Implementation

5.2   Experiments

In order to evaluate our prototype implementation, we set up a small testbed, containing a mobile host and a remote correspondent host, both running our prototype M-SCTP implementation. The two nodes where connected via a WAN emulator node in order to introduce some delay, and the mobile node is connected to this WAN emulator (NIST Net) via two network interfaces which can individually be turned on or off, emulating the obtaining or loosing of contact with a base station. The WAN emulator was set to provide a delay of 80 ms and 10 MBit/s bandwidth. The mobile node continuously sends packets to the correspondent node. This setup is outlined in Figure 3.

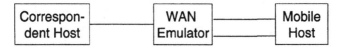

Figure 3.   Network topology used for M-SCTP experiments

5.3   Result and analysis

The first result of the prototype implementation is that it is indeed possible to support handoff based on SCTP's multi-homing mechanisms. As a first performance evaluation, we were interested in the resulting interruption time. Figure 4 shows a sequence number trace over time. At point $t = 8$, the old network interface is turned off, and it takes about one second until new packets arrive. This long delay is due to the fact that our prototype is still using the original SCTP path management mechanisms. These mechanisms are tuned for fixed-network error handling, not for supporting mobility. Nevertheless, replacing these mechanisms with better solutions based on link-layer information is clearly possible and one objective of current work.

6.   Conclusion and Future Work

Solving the IP mobility problem in an end-to-end fashion is possible. A transport layer protocol that provides multi-homing-like mechanisms as SCTP does is a natural candidate for supporting such a mobility approach. Additionally, leveraging multi-homing opens the possibility to exploit simultaneous connectivity to multiple access points in a flexible and straightforward fashion.

The main reason for prototype's somewhat long interruption time is mostly due to its reliance on SCTP's path management mechanisms: these must be replaced in order to realize the full potential of our architectural approach. This is the focus of current work and, in particular, the integration with link-layer triggering should provide considerable improvements over the standard case.

146

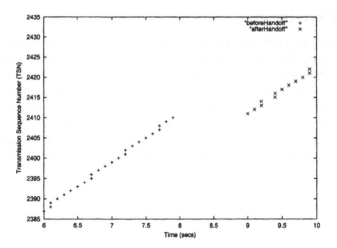

Figure 4. An SCTP association sequence trace during handoff

When to actually start the registration of a new IP address with the correspondent host is a further issue.

Finally, the behavior of M-SCTP in the case that both end points are mobile hosts is not clear when both are moving simultaneously. The immediately obvious solution for this problem is to use dynamic DNS mechanisms, but as this is a rather heavy-weight solution, there should be some potential for additional improvements here.

Acknowledgments

The authors would like to thank their colleagues from the TKN Group, especially Xiaoming Fu and Hongbong Kim. Part of this work was supported by a research contract with Siemens AG, Munich, Germany.

References

[1] R. Droms. Dynamic host configuration protocol. RFC 1531, October 1993.

[2] C. E. Perkins. IP Mobility Support. RFC2002, October 1996.

[3] M. Riegel and M. Tuexen. Mobile SCTP. Internet draft draft-riegel-tuexen-mobile-sctp-00.txt (work in progress), February 2002.

[4] J. H. Saltzer, D. P. Reed, and D. D. Clark. End-to-end arguments in system design. *ACM Trans. on Computer Systems*, 2(4):277–288, 1984.

[5] A. C. Snoeren and H. Balakrishnan. An end-to-end approach to host mobility. In *Proc. 6th ACM/IEEE Intl. Conf. on Mobile Computing and Networking (MobiCom)*, Boston, MA, August 2000.

[6] R. Stewart. Stream Control Transmission Protocol. RFC2960, November 2000.

[7] R. Stewart and Q. Xie. SCTP extensions for Dynamic Reconfiguration of IP Addresses. Internet draft draft-ietf-tsvwg-addip-sctp-04.txt (work in progress), January 2002.

GRAPH BASED REPRESENTATION OF OBJECTS IN IMAGE RETRIEVAL

S. Bischoff

Heinrich-Hertz-Institute for Communication Technology
Einsteinufer 37, 10587 Berlin, Germany
bischoff@hhi.de

F. Wysotzki

Technical University Berlin Institute of Artificial Intelligence
Franklinstr. 28/29, 10587 Berlin, Germany
wysotzki@cs.tu-berlin.de

Abstract   A new distance measure to compare whole and parts of images is proposed. This measure considers the color, shape and texture properties of image segments as well as their relative mutual positions.

Keywords: Image Retrieval, Image Segmentation, Image Indexing, MPEG-7 Descriptors, Graph Metric, Maximum Weight Clique Problem, Hopfield-Style Neural Network, WTA-Net

1.   Introduction

Retrieval of digital images based on image features rather than text descriptions has been subject of considerable research efforts over the last years. Within this framework the MPEG-7 standard (Moving Picture Experts Group) attempts to standardize suitable descriptors to allow many image search engines to access as many distributed images as possible. Various image descriptors are currently under consideration in the MPEG video group, such as color and color distribution descriptors, texture and shape descriptors (Meiers et al., 2001).

Unfortunately, most descriptors are integral quantities and depend strongly on the integration area. Calculation on the whole image or on bad areas may misrepresent them. In addition, the typical user searches for images in which semantical objects like cars, houses or persons occur. That cannot be described by feature vectors calculated from the whole

G. Hommel and S. Huanye (eds.), The Internet Challenge: Technology and Applications, 149–157.

150

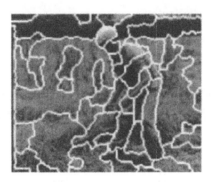

Figure 1. Segmented Image

image. An adequat description rather needs both: a suitable representation of the visual properties of semantic image parts and their mutual positions, too.

In the following we describe our approach to transform an image into a labeled graph. The nodes present features of the segments and the edges relative positions, respectively. This graph is then the basis of a fast recognition procedure using a special Neuronal Net for computing a graph distance.

2. Image Segmentation

In most computer vision applications image segmentation constitutes a crucial initial step before performing high-level tasks such as object recognition and scene interpretation. Automatic image segmentation is well known as one of the hardest low-level vision problems. It is not guaranteed that a coherent cluster in feature space corresponds to a coherent image region. Several algorithms are known like (Fuh et al., 2000; Weickert, 1998; Winter and Nastar, 1999). We chose "Edge Flow" (Ma and Manjunath, 1997), which utilizes a predictive coding model to identify the direction of change in color and texture at each image location at a given scale, and constructs an edge flow vector. By iteratively propagating the edge flow, the boundaries can be detected. After boundary detection, disjoint boundaries are connected to form closed contours.

Today, no system is able to segment any image into semantic meaningful parts without additional information. Furthermore the segmentation result depends strongly on illumination conditions, camera position and algorithm parameters.

In the next section we describe which visual properties are extracted for each segment.

3. Extraction of MPEG-7 Descriptors

Scalable Color Descriptor (SCD): The Scalable Color Descriptor describes the color distribution in the segment. The MPEG-7 generic SCD is a color histogram in HSV colorspace. The color histogram is based on the definition of uniform colorspace quantization with 16 bins in hue and 4 bins in both saturation and value.

Homogenous Texture Descriptor (HTD): The Homogenous Texture Descriptor describes directionality, coarseness and regularity of patterns in the segment. In order to describe the segment texture, energy and energy deviation values are extracted from a frequency layout. Suitable descriptions are obtained in the frequency domain by computing mean and standard variation of frequency coefficients (Sikora, 2001). To handle the dependence on the segment shape, only higher frequencies were taken into account. The descriptor is based on convolution with a filter bank consisting of 12 2d Gabor wavelets (3 scales and 4 directions). This can be quickly done by filtering with fourier transformed 2d Gabor wavelets in the fourier domain.

Edge Histogram Descriptor (EHD): Edge Histogram Descriptor represents the spatial distribution of five types of edges. The edge types are a horizontal, a vertical, two different diagonal and a non-directional edge type. So the edge histogram with 5 bins forms a feature vector with 5 components. For more details see (Meiers et al., 2001) and (Sikora, 2001).

4. Image Representation by Labeled Graphs

Labeled graphs are an appropriate and popular representation of structured objects in many domains. To transform an image into a labelled graph the centroid for each segment was calculated. The set of all centroids is the set of nodes. Every node is labeled with an MPEG-7 feature vector. Edges, labeled with distances between centroids, are introduced for each pair of nodes. The distances are normalized with respect to the image diagonal. The Neighbourhood relation is one additional component of the edge feature vector. So the RAG (Region Adjacency Graph) is included.

5. Image Distance

To compare two images in the first step graph normalization is computed for both RAG1 and RAG2. In the second step computation of a graph metric between the normalized RAG1 and RAG2 defines the image distance. We describe the first step in the next section.

5.1 Graph Normalization

Images are considered to be similar, if the distance between their representation by a configuration of features is small. Caused by the graph representation used here it is necessary to compute a graph distance. Most graph distances need a unique node assignment. But the segmentation results depend strongly on illumination conditions, camera position and algorithm parameters. Different graph representations of one image are possible. So we try to unify such neighbouring segments in the first image and vice versa, which correspond to one segment in the second one and which have suitable topological relations to all other segments in the first image.

For this purpose let be $G_1 := (N_{G_1}, V_{G_1}, l_{G_1}, e_{G_1}, \theta_{G_1})$ the Region Adjacency Graph of image 1 and $G_2 := (N_{G_2}, V_{G_2}, l_{G_2}, e_{G_2}, \theta_{G_2})$ the RAG of image 2. G_1 and G_2 are two complete undirected graphs with finite node sets N_{G_1}, N_{G_2} and (finite) edge sets $V_{G_1} = N_{G_1} \times N_{G_1}$, $V_{G_2} = N_{G_2} \times N_{G_2}$. With $l_{G_i} : N_{G_i} \to \mathbb{R}^n$ (MPEG-7 feature vector) and $e_{G_i} : V_{G_i} \to \mathbb{R}$ (normalized euclidean distance) we denote the node and edge label functions for $i \in \{1,2\}$, respectively. Let be $\sigma_L : \mathbb{R}^n \times \mathbb{R}^n \to [0,1]$ and $\sigma_E : \mathbb{R} \times \mathbb{R} \to [0,1]$ mappings, which measure the similarity of nodes and edges respectively ($[0,1] \simeq$ [not similar,identical]). With θ we describe the adjacency relation. If x_1 and x_2 are adjacent, then $(x_1, x_2) \in \theta$. By definition always $(x, x) \in \theta$, too. Let be F, G the set of mappings

$$F := \{f : N_{G_1} \to N_{G_2} \mid f(x_i) = f(x_j) \Rightarrow (x_i, x_j) \in \theta \}$$

$$G := \{g : N_{G_2} \to N_{G_1} \mid g(y_i) = g(y_j) \Rightarrow (y_i, y_j) \in \theta \}$$

and

$$Q(f) := \sum_{x \in N_{G_1}} \sigma_L(l_{G_1}(x), l_{G_2}(f(x))) + \sum_{(x_i, x_j) \in V_{G_1}} \sigma_E(e_{G_1}(x_i, x_j), e_{G_2}(f(x_i), f(x_j)))$$

$$Q(g) := \sum_{x \in N_{G_1}} \sigma_L(l_{G_2}(y), l_{G_1}(g(y))) + \sum_{(x_i, x_j) \in V_{G_2}} \sigma_E(e_{G_2}(y_i, y_j), e_{G_1}(g(y_i), g(y_j))).$$

We unify 2 segments x_i and x_j of image 1, if $(x_i, x_j) \in \theta$ and $f^\star(x_i) = f^\star(x_j)$ with $f^\star := \arg \max_{f \in F} Q(f)$. For $(y_i, y_j) \in \theta$ we do the same, if $g^\star(y_i) = g^\star(y_j)$ and $g^\star := \arg \max_{g \in G} Q(g)$. Now we calculate a new RAG1 and RAG2 and iterate, until no more segments can be unified. We appply a connectionistic method to compute $f^\star$ and $g^\star$. This connectionistic method is declared in next section in detail to determine a best mapping $\Phi^\star$ from G_1 into G_2. Computation of $f^\star$ and $g^\star$ is similar.

5.2 A new Graph Metric and its Computation

Provided a one to one match (unique node assignment) we generalize and proof a metric property for abstract graphs with node label set L and egde label set E. In the context of image retrieval there is $L = \mathbb{R}^n$ and $E = \mathbb{R}$.

Definition 5.1 *A mapping* $\Phi : N_{G_1} \to N_{G_2}$ *is called generalized graph isomorphism* $:\Longleftrightarrow$

1 Φ *is without consideration of node and edge labels a graph isomorphism.*

2 $\forall x_i, x_j \in N_{G_1} : \sigma_L(l_{G_1}(x_i), l_{G_2}(\Phi(x_i))) > 0 \wedge$
$\sigma_E(e_{G_1}(x_i, x_j), e_{G_2}(\Phi(x_i), \Phi(x_j))) > 0$.

Then G_1 *and* G_2 *are called g-isomorph.*

Definition 5.2 *A mapping* $\Phi : N_{G_1} \to N_{G_2}$ *is called generalized subgraph isomorphism* $:\Longleftrightarrow \exists$ *an induced subgraph* G_1' *of* G_1 *and* $\exists$ *an induced subgraph* G_2' *of* G_2 *and* $\Phi \mid_{N_{G_1'}} : N_{G_1'} \to N_{G_2'}$ *is a generalized isomorphism.*

Definition 5.3 *Let* Φ *be a generalized subgraph isomorphism.* G_1' *and* G_2' *being the g-isomorphic induced subgraphs. We define as* support of Φ $supp\Phi := \{x \in N_{G_1} : x \in N_{G_1'}\}$ *the set of all nodes of* G_1 *mapped g-isomorph.*

Now consider the following functional d which measures the distance of two labeled graphs G_1 and G_2:

$$d(G_1, G_2) := \min_{\{\Phi \, : \, \Phi \, {g-subgraph- \atop isomorphism}\}} (\mu(\mu - 1) - v_\Phi) + \gamma(\mu - n_\Phi)$$

with

$$\mu := \max\{\mid N_{G_1} \mid, \mid N_{G_2} \mid\}$$
$$n_\Phi := \sum_{x \in \text{supp}\Phi} \sigma_L(l_{G_1}(x), l_{G_2}(\Phi(x)))$$
$$v_\Phi := \sum_{(x_i, x_j) \in (\text{supp}\Phi)^2} \sigma_E(e_{G_1}(x_i, x_j), e_{G_2}(\Phi(x_i), \Phi(x_j)))$$

and $\gamma \in [0, 1]$ measuring the relative contribution of edges and nodes, respectively. This function is well defined except for nodes and edges with similarity zero. Then the following proposition holds:

Proposition 5.1 *d is a metric in the space of isomorphic classes of labeled graphs with the same label sets L, E.*

Proof: (See Schaedler and Wysotzki, 1999)

To quantify d, it is necessary to specify the best mapping $\Phi^* = \arg\max(v_\Phi + \gamma n_\Phi)$ among all possible g-subgraph isomorphisms from G_1 into G_2. If we enumerate all nodes of G_1 and G_2, every mapping from G_1 into G_2 can be described by a matrix $o \in \{0,1\}^{|N_{G_1}| \times |N_{G_2}|}$ where $o_{ik} = 1$ if $\Phi(x_i) = y_k$ and $o_{ik} = 0$ else. So we can formulate this problem as an optimization problem:

$$o^* = \arg\max_o \left(\sum_{i,j,k,l} o_{ik} s_{ij,kl} o_{jl} + \gamma \sum_{i,k} o_{ik} s_{ik} \right)$$

with $s_{ij,kl} = \sigma_E(e_{G_1}(x_i, x_j), e_{G_2}(y_k, y_l))$ and $s_{ik} = \sigma_L(l_{G_1}(x_i), l_{G_2}(y_k))$ with uniqueness restriction $o_{ik} o_{il} = o_{ik} o_{jk} = 0 \ \forall i \neq j$, $k \neq l$ and induced subgraph restriction $\prod_{i,j,k,l} s_{ij,kl} \neq 0$. These restrictions can be included by means of a penalty term *pen*. So we can rewrite with minimal node and edge similarities Θ_N, Θ_V:

$$o^* = \arg\max_o \left(\sum_{i,j,k,l} o_{ik} \hat{s}_{ij,kl} o_{jl} + \gamma \sum_{i,k} o_{ik} \hat{s}_{ik} \right)$$

with

$$\hat{s}_{ik} = \begin{cases} s_{ik} & s_{ik} \geq \Theta_N \\ 0 & \text{else} \end{cases}$$

and

$$\hat{s}_{ij,kl} = \begin{cases} \frac{s_{ij,kl} + s_{ji,lk}}{2} & i \neq j \wedge k \neq l \wedge \frac{s_{ij,kl} + s_{ji,lk}}{2} \geq \Theta_V \\ -pen & i = j \vee k = l \\ -pen & \frac{s_{ij,kl} + s_{ji,lk}}{2} < \Theta_V \end{cases}$$

Solving this optimization problem is equal to find this common generalized induced subgraph with maximum summarized nodes and edges similarity. This can be regarded as maximum weight clique problem in the generalized compatibility graph (ECG), which will be described now.

Definition 5.4 *The generalized compatibility graph $ECG = (N, V, l, e)$ of two graphs $G_1 = (N_{G_1}, V_{G_1}, l_{G_1}, e_{G_1})$ and $G_2 = (N_{G_2}, V_{G_2}, l_{G_2}, e_{G_2})$ is constructed as follows:*

1 $N := \{(x_i, y_k) : x_i \in N_{G_1} \wedge y_k \in N_{G_2} \wedge \sigma_L(l_{G_1}(x_i), l_{G_2}(y_k)) \geq \Theta_N\}$

2 $l((x_i, y_k)) := \hat{s}_{ik}$

3 $V := N \times N$

4 $e((x_i, y_k), (x_j, y_l)) := \hat{s}_{ij,kl}$

The maximum weight clique problem is a generalization of the maximum clique problem which is well known to be NP-complete. To solve this problem we use a Winner-Takes-All Net (WTA-Net), which is a Hopfield style Neural Network with excitatory and inhibitory connections. The WTA-Net is constructed directly from ECG by using the following rules:

1 For every node $(x_i, y_k) \in N$ with label $\hat{s}_{ik}$ create a unit $u(x_i, y_k)$ of the net.

2 For every edge $((x_i, y_k), (x_j, y_l)) \in V$ with positive label $\hat{s}_{ij,kl}$ create a (symmetric) connection $(u(x_i, y_k), u'(x_j, y_l))$ with positive initial weight $\omega_{ij,kl} \sim \hat{s}_{ij,kl}$ (excitatory connection).

3 For every edge $((x_i, y_k), (x_j, y_l)) \in V$ with the label -pen create a (symmetric) connection $(u(x_i, y_k), u'(x_j, y_l))$ with negative initial weight $\omega_{ij,kl} \sim -pen$ (inhibitory connection).

4 Every unit $u(x_i, y_k)$ gets a bias input $I_{ik} \sim \hat{s}_{ik}$.

The units are updated synchronously, according to the rule:

$$p_{ik}(t+1) = (1-d)p_{ik}(t) + I_{ik} + \sum_{(j,l) \neq (i,k)} \omega_{ij,kl} o_{jl}(t)$$

$$o_{ik}(t+1) = \begin{cases} 0 & p_{ik}(t+1) < 0 \\ p_{ik}(t+1) & p_{ik}(t+1) \in [0,1] \\ 1 & p_{ik}(t+1) > 1 \end{cases}$$

A steady state of the net defines a mapping between the two original graphs. The distance measure d can be calculated based on this mapping.

6.   Results

Image retrieval makes great demands on fast matching as well as suitable retrieval results. Taking color histograms for fixed subimages is a way often used in exploiting spatial layout (Meiers et al., 2001). The retrieval is fast, but the results are not so satisfactory (Figure 2). There are images retrieved which exhibit no similarity with the query image. Considering the relative positions of regions is more sophisticated (Carson et al., 1999; Fuh et al., 2000). Fuh, Cho and Essig use a combination of color segmentation with a relationship tree and a corresponding tree-matching method (Fuh et al., 2000). This information-reduced representation of images allows fast retrieval, because powerful mathematical substring matching methods can be applied using dynamic programing. The disadvantage of this method is that tree structures must be the same in query and database image. If the position of segments changes a little, this requirement may not be fulfilled.

Our approach considers the visual properties of image segments as well as their relative mutual positions. The user has the choice to search for a whole image or only for special image parts. Figure 3 shows a segmented query image. Segments which are searched for are outlined black. For the corresponding labeled graph, graph normalization and distance measure d to all other graph representations of database images are calculated. Images with shortest distances are shown in Figure 4.

The computation time depends on the order of graph representations and quantity of database images. A request for Figure 3 for instance needs 33 seconds computation time on P2/333MHz and 100 database images.

References

Carson, C., Thomas, M., Belongie, S., Hellerstein, J., and Malik, J. (1999). Blobworld: A system for region-based image indexing and retrieval. In *Proceedings of the Third International Conference VISUAL '99, Amsterdam, The Netherlands, June 1999*, Lecture Notes in Computer Science 1614. Springer.

Fuh, C., Cho, S., and Essig, K. (January 2000). Hierarchical color image region segmentation for content based image retrieval system. *IEEE TRANSACTIONS ON IMAGE PROCESSING*, 9(1).

Ma, W. and Manjunath, B. (1997). Edge flow: a framework of boundary detection and image segmentation. *Proc. of CVPR*.

Meiers, T., Keller, I., and Sikora, T. (May 2001). Image visualization and navigation based on MPEG-7 descriptors. *Conference on Augmented, Virtual Environments and 3D Imaging, EUROIMAGE 2001*.

Schaedler, K. and Wysotzki, F. (1999). Comparing structures using a hopfield-style neural network. *Applied Intelligence*, 11:15–30.

Sikora, T. (June 2001). The MPEG-7 visual standard for content description. *IEEE TRANSACTIONS ON CIRCUITS AND SYSTEMS FOR VIDEO TECHNOLOGY*, 11(6).

Weickert, J. (1998). *Mustererkennung 1998*, chapter Fast segmentation methods based on partial differential equations and the watershed transform, pages 93–100. Springer Verlag.

Winter, A. and Nastar, C., editors (June 1999). *Differential Feature Distribution Maps for Image Segmentation and Region Queries in Image Databases*, Ft Collins, Colorado. IEEE CVPR'99 workshop on Content Based Access of Images and Video Libraries.

Figure 2. Classical search request. First image is query image.

Figure 3. Segments which are searched for are outlined black.

Figure 4. Result of search request for our approach. First image is query image.

SPATIAL INFERENCE AND CONSTRAINT SOLVING

How to Depict Textual Spatial Descriptions from Internet

Carsten Gips
Berlin University of Technology
cagi@cs.tu-berlin.de

Fritz Wysotzki
Berlin University of Technology
wysotzki@cs.tu-berlin.de

Abstract   Today there are still many applications in the Internet, where the user is given a textual description of a spatial configuration (e.g. chat, e-mail or newsgroups). The user is asked to imagine the scene and to draw inferences. We present a new approach to generate depictions of such scenes. Besides of drawing spatial inferences, this leads to the problem of solving a system of complicated numerical constraints. In contrast to qualitative spatial reasoning, we use a metric description where relations between pairs of objects are represented by parameterized homogenous transformation matrices with numerical (nonlinear) constraints. We employ methods of machine learning in combination with a new algorithm for generating depictions from text including spatial inference.

Keywords: Spatial Reasoning, Constraint Satisfaction, Machine Learning, Depictions

1.   Introduction

There are many fields where it is important to understand and interpret textual descriptions of real world scenes. Examples are navigation and route descriptions in robotics (Röfer, 1997; Jörding and Wachsmuth, 1996), CAD and graphical user interfaces (e.g. "The xterm is right of the emacs.") or visualization of scenes given in the Internet (e.g. in newsgroups or in e-mail).

In contrast to qualitative approaches to spatial reasoning (Allen, 1983; Guesgen, 1989; Hernández, 1994), we presented in Claus et al., 1998 a new metric approach to spatial inference based on mental models (Johnson-Laird, 1983). Starting from textual descriptions containing sentences like "The lamp is left of

159

G. Hommel and S. Huanye (eds.), The Internet Challenge: Technology and Applications, 159–168.
© 2002 *Kluwer Academic Publishers.*

the fridge." we try to construct a mental model which represents the described spatial situation. This approach uses a directed graph, where the nodes represent the objects and the edges represent the given relation between two objects, e.g. left(fridge, lamp). From this model it is possible to infer relations which were not initially given in the text (i.e. to answer questions about the described spatial scene or to complete the model). In a further step we can use the model to generate depictions compatible with the description.

The semantics of the relations is given by homogenous transformation matrices with constraints on the variables. As shown in Wiebrock et al., 2000, inference of a relation between two objects is done by searching a path between the objects and multiplying the matrices on this path. Thereby constraints containing inequalities and trigonometric functions must be propagated and verified. Only in some rare cases we can solve these constraints analytically. Wiebrock et al., 2000 proposed a simple algorithm for generating depictions. It is restricted to default positions of objects and to rotations of multiples of $\pi/2$. Moreover, they had to keep lists with possible positions for every object.

Our aim is to find a method to solve this kind of constraints and to generate depictions without these restrictions. We sketch an approach to spatial reasoning which applies machine learning in combination with a new algorithm for depiction generation.

This paper is structured as follows: We start with an introduction into the description of spatial relations in Sect. 2. In Sect. 3 we apply methods of machine learning to learn the semantics of the relations, in order to obtain an alternative representation of the constraints. Afterwards we sketch in Sect. 4 a new approach for generating depictions of the described spatial layout, i.e. solving the constraints, which uses the results of the machine learning step. At the end of the article we give in Sect. 5 a conclusion and draw some research perspectives.

2. Expressing Spatial Relations

Starting from texts with descriptions of spatial layouts, we want to generate appropriate depictions. Additionally we have to determine whether the given descriptions are inconsistent, i.e. whether there are no possible depictions.

The texts describe the scenes by the use of spatial relations. We investigated scene descriptions based on the relations left/2 and right/2, which describe the placement of an object left resp. right of another one, the relations front/2 and behind/2 which place objects in front of or behind other objects, resp., and the relation atwall/2 for describing the placement of an object parallel to a wall with a fixed maximum distance. Further relations provide background knowledge (i.e. an object is always situated in a given room and the objects must not overlap). For simplification, we consider 2D scenes only and represent objects by appropriate geometric figures.

As mentioned above, in contrast to qualitative techniques (Allen, 1983; Guesgen, 1989; Hernández, 1994) we use a metric approach for spatial reasoning (Claus et al., 1998; Geibel et al., 1998) known from the area of robotics (Ambler and Popplestone, 1975). We associate with every object a coordinate system, and its form and size. Relations between pairs of objects are represented by constraints on parameters of their transformation matrices. Thus, the current coordinates of an object are expressed relative to the coordinate system of its relatum, which may be different in different relations. That means, when changing the relatum of an object, we need to transform the coordinates of this object by multiplying them with the corresponding matrix.

Let us consider the relation right/2 in detail. Called with cupboard and lamp as arguments (right(cupboard, lamp)), it places the lamp, which is the referent, right wrt. its relatum, the cupboard. Thus the coordinate system of the cupboard is the origin of the relation. The lamp can be placed in the area restricted by the upper and lower bisectors of the upper and lower right corner and the right side of the rectangle. Figure 1 illustrates this situation.

Mathematically we can describe the relation $right(O1, O2)$[1] by the inequalities 1a, 1b, and 1c.

$$\Delta x_2{}^1 \geq O1.w + O2.r \tag{1a}$$
$$\Delta x_2{}^1 \geq \Delta y_2{}^1 + O1.w - O1.d + \sqrt{2}\,O2.r \tag{1b}$$
$$\Delta x_2{}^1 \geq -\Delta y_2{}^1 + O1.w - O1.d + \sqrt{2}\,O2.r \tag{1c}$$

Thereby $O1.w$ and $O1.d$ represent the width and the depth of the rectangle, i.e. the cupboard, and $O2.r$ stands for the radius of the lamp. The distances of the object $O2$ in the x- and y-directions from the relatum $O1$ are denoted by $\Delta x_2{}^1$ and $\Delta y_2{}^1$, resp. Each object has intervals for its variables, because it stands for a class of objects (i.e. the object cupboard is a "cupboard frame", standing not only for a particular cupboard but for all possible cupboards).

Note, that for the relation right/2, like for every spatial relation, the formulae differ generally depending on the form of the relata and referents.

3.   Learning Spatial Relations with CAL5

In our problem domain (objects to be located in a room), the constraints consist of equations and inequalities containing trigonometric functions which lead to computational difficulties well known from robotics (Ambler and Popplestone, 1975).

Instead of solving the constraints directly we try to learn the decision function $C(x_1, \ldots, x_n)$ which decides whether a vector $x = (x_1, \ldots, x_n)$ of the

---

[1]$O1$ and $O2$ may stand for the cupboard and the lamp, resp.

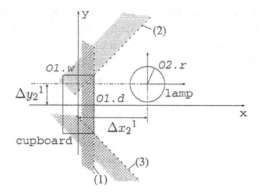

Figure 1.   The relation right(cupboard, lamp) in detail

configuration space belongs to a region where the predicate C is true, i.e. the corresponding constraints are satisfied.

Before employing machine learning algorithms we have to construct a training set by exploiting the given constraint description or by using results of psychological studies. These datasets consist of preclassified feature vectors where each variable of the constraints represents an attribute, i.e. a dimension in the feature space. In the following, we will use "class A" for the regions where a constraint C is satisfied (and "class B" otherwise). By means of the training sets, algorithms of classification learning (e.g. decision tree learning like CAL5 or neuronal nets like Dipol, see Müller and Wysotzki, 1997 or Schulmeister and Wysotzki, 1994, resp.) construct classifiers. These decide the class membership of an arbitrarily chosen point x (not necessarily contained in the training set) by inductive generalization. Generating training sets in order to get an acceptable approximation of the decision boundary is also known as "learning by exploration" or "active learning" in literature.

In this work we have chosen CAL5 for learning the spatial relations. CAL5 approximates, as it is a decision tree learner, the class boundaries piecewise linearly by axis-parallel hyperplanes. Usually, there is a generalization error due to the unavoidable approximation of the boundaries between the A-regions and the B-regions. This error can be measured using a test set of classified example vectors different from the training set. By increasing the number of training data (and by simultaneously shrinking a certain parameter of CAL5) the generalization error can be reduced (i.e. the accuracy of the class boundary approximation can be made arbitrarily high), and in the limit of an infinite set of training data the error becomes zero. In Geibel et al., 1998 we investigated the problem "a bar is right of an object O", represented in the configuration space defined by the angle of the bar with the x-axis and the displacement of the bar with respect to the origin S of the coordinate system of the relatum O. We also demonstrated experimentally, that the generalization error of the obtained decision tree shrinks with an increasing number of points for learning. However, in practice we reach the manageable limit at 200.000 training examples. The constraints of our relations (like right/2 for circles and rectangles, see Sec. 2) affect up to seven parameters, thus, we obtain a configuration space with up to seven dimensions. This corresponds to approximatively ten data points per dimension[2]. Because of this sparsely populated configuration space both the training and generalization errors are rather high. This is shown in Tab. 1, where we used 200.000 uniformly distributed data points for learning and 5.000 points for testing.

Relation[3]	Number of class A leafs	Points in A	in B	Test error for A only	for B only	total
atwall(r, r)	143	15.909	184.091	10 %	1 %	2 %
front(c, r)	1.984	84.285	115.715	22 %	14 %	17 %
right(c, r)	1.702	84.557	115.443	20 %	13 %	16 %
right(r, c)	2.079	87.550	112.450	26 %	14 %	19 %

Table 1. Results of the learning process for some spatial relations

Furthermore we have to be aware of generating points in a sufficient large subspace of configuration space. Figure 2 shows the result of a too small scope of the training data. The resulting classifier does not cover the intersection of the right/2 sector with the room area.

The benefits of our learning approach are to get a new, easier representation of the decision boundary (i.e. the constraints). The new representation contains the solution of the constraints (i.e. the A-regions), and the accuracy of the approximation can be made arbitrarily high. The problem, however, is the generation of suitable data sets.

4. Generating Depictions

In the previous section we transformed the constraints of the spatial relations into a new representation (i.e. the learned classifiers). These CAL5 decision

---

[2]Consider the relation right/2 for two rectangles. This problem has seven parameters, so the configuration space consists of seven dimensions. Supposed we obtain the same number of data on each dimension (like a grid), the 7th root of 200.000 yields approximately six data points on each dimension. Note, this is not a correct calculation but a simple estimation.

[3]r denotes a rectangle, and c a circle, resp.

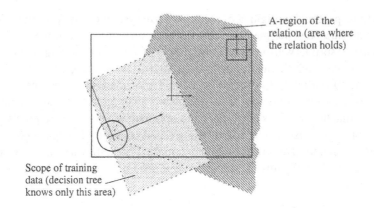

A-region of the relation (area where the relation holds)

Scope of training data (decision tree knows only this area)

Figure 2. Scope of training data not adapted to the admeasurement of the room, the classifier may not cover the intersection of the right/2 sector with the room area

trees are used by a new algorithm (see Fig. 3) for generating depictions, i.e. for simultaneously solving the set of given spatial constraints.

As mentioned above, for every needed relation and for every pair of object types, corresponding (sub-)relations must be learned. Recall that our relations are binary. Thus, we get three cases: both objects are 'unknown', one object is already placed or both objects are placed. In the first case, we place one object randomly in the room (line 11 in Fig. 3). This leads us to the second case. There we pick a class A leaf of the tree and compute the size, relative position and relative orientation of the other object by assigning values to the remaining variables of this object within the intervals of the chosen leaf (line 12 and 13). If the collision check in line 14 fails (for both, case one and two), we repeat the procedure up to k times. If we do not have admissible values after the kth trial, we suppose that the current relation cannot hold in combination with the others and reject the depiction generated so far. However, there may exist solutions. Actually we cannot distinguish between the case "no solution" and "disadvantageous values". So if we reject the vector[4], we have to start again with the first relation. For practical reasons we work instead on a number of object constellations[5] in parallel. In the last case both objects are already placed and we have to check, wether the values of the objects range in the intervals represented by at least one leaf (line 05 to 07). If not, the relations do not hold, at least for the calculated values.

---

[4]Note, that each depiction is a point in the configuration space. This points are represented by vectors.
[5]Initially we have a number v of empty vectors.

Depiction generation algorithm

INPUT:

number v of initial vectors and number k of trials

relations r that have to hold

OUTPUT:

up to v depictions, where all relations r hold

ALGORITHM:

```
01   foreach relation r:
02       identify objects and object types by object descriptions
03       load the corresponding decision tree
04       foreach vector v:
05           if both objects were placed
06           then
07              check whether relation r holds
08           else
09              if both objects are new
10              then
11                 place first object randomly in room
12              pick randomly class A leaf
13              assign values to variables within intervals of leaf
14              check non-overlapping with other objects and walls
15              repeat up to k times if check fails
16           if no success
17           then
18              drop vector v
19   show remaining vectors (depictions)
```

Figure 3.   Algorithm for generation depictions using decision trees

This procedure is repeated for every relation with the remaining objects, which satisfy the relations processed in the former steps. Finally we obtain up to v depictions according to the given spatial description. In the case, that we have not found any depiction, we have to assume that the constraints are unsatisfiable.

Up to now each decision tree represents only the constraints for the particular relations with the two objects of the specified forms. The background knowledge[6] was not learned but will be checked after every step explicitly. In

---

[6]E.g. the objects must not overlap.

general, it is possible to learn the background knowledge constraints as well and to check them like the other relations.

The input parameters v and k depend on the number and on the type of the given relations. They have to be chosen large enough to get a correct answer ("There is no solution." or "We have found at least one.") with some probability. At the same time, one should choose rather small v and k, because by increasing the values of the parameters the calculation time increases, too. So they have to be chosen in relation to the problem to solve. As shown in Tab. 2, the more relations to solve, the higher v has to be.[7] A value of 100 for k seems to be a good choice. The number of trials per valid solution increases exponentially in the number of relations to be solved.

Not shown, but critical is the processing sequence. Restrictive relations like atwall/2 should be solved at the beginning. Supposed we have two relations, right(cupboard, lamp) and atwall(wall1, cupboard). Now we fulfill first the right/2 relation. Therefore the cupboard may be placed somewhere in a relatively large area in the room. After that we may be unable to satisfy atwall/2, just because the cupboard should be placed nearby wall1, but actually it is already placed at another area in the room. So we would have to increase v, but nevertheless the probability to get a solution is very small.

5. Discussion

In the previous sections we sketched a new approach to solve the constraints occuring in spatial reasoning. Instead of solving the constraints directly, we employed methods of machine learning for transforming the constraints into another representation. Using the decision tree learner CAL5 yields interpretable results. The approximation of the decision boundaries may be (at least in principle) arbitrarily high. Generating suitable training sets, however, is not trivial and is subject of current research ("active learning", Wiebrock and Wysotzki, 1999). So we have to deal with an increasing amount of disk space and calculation time and have to care for a suitable distribution of the training data.

After learning we have, due to the obtained decision trees, detailed knowledge about the regions in the configuration space. The depiction generation algorithm employs the decision rules for restricting the space to find possible solutions. In the limit of generating an infinite number of depictions (i.e. exhaustive search) the algorithm finds every possible solution. Because the processing sequence of the relations is critical we may find no solution, although there is one. However, the scene descriptions in this problem domain are usually underconstrained, and, thus, it is usually not a problem to find an alternative solution.

---

[7]For testing we implemented a first prototype in Perl, which is quite slow in comparison to usual programming languages, like C or Java. So we forgo to show the running times.

Combination of relations	number v of initial vectors per valid depiction (average)
single relation, e.g. right(steffi, cupboard)	2
two relations, e.g. right(steffi, cupboard) front(steffi, fridge)	10
three relations, e.g. right(steffi, cupboard) left(fridge, lamp) right(cupboard, lamp)	61
four relations, e.g. right(steffi, cupboard) left(fridge, lamp) right(cupboard, lamp) front(steffi, fridge)	375

Table 2. Some test runs and typical results of our algorithm

by constructing another sequence (i.e. following another path in the problem space).

This yields further research perspectives: First of all, we could use traditional constraint solving systems, like Hofstedt, 2000, for restricting the search space and for reducing the training cost. They could be employed for precomputing to get a sure exclusion of unsatisfiable scene descriptions (Gips et al., 2002). However, because of the incompleteness of the solvers, they will not detect all inconsistencies. Instead they yield regions, which are too large, but include the searched solution areas. So we could use the results of the constraint solvers for generating better training sets with fewer data points. Secondly, we have to investigate strategies for selecting a particular leaf of a decision tree in the depiction algorithm. Until now we use the volume of the region described by a leaf for selecting a particular leaf for further calculation. Recent studies (Wiebrock, 2000) have shown that a better selection criterion could dramatically improve the speed. Furthermore, we have to deal with the processing sequence of the relations. As explained, very restrictive relations should be satisfied first. On the basis of the decision trees we have to develop a measure for the restrictiveness. Thirdly, we should consider the principle of the least astonishment. We visualize spatial descriptions and draw inferences on the behalf of the user. So we should return those depiction (if there is any), which the user is most accustomed to. This could be achieved by using specific distributions for the training data, as first results in Wiebrock, 2000 show.

References

Allen, J. F. (1983). Maintaining Knowledge about Temporal Intervals. *Comm. of the ACM*, 26:832–843.

Ambler, A. P. and Popplestone, R. J. (1975). Inferring the Positions of Bodies from Specified Spatial Relationships. *Artificial Intelligence*, 6:157–174.

Claus, B., Eyferth, K., Gips, C., Hörnig, R., Schmid, U., Wiebrock, S., and Wysotzki, F. (1998). Reference Frames for Spatial Inferences in Text Comprehension. In Freksa, C., Habel, C., and Wender, K. F., editors, *Spatial Cognition*, volume 1404 of *LNAI*. Springer.

Geibel, P., Gips, C., Wiebrock, S., and Wysotzki, F. (1998). Learning Spatial Relations with CAL5 and TRITOP. Technical Report 98-7, TU Berlin.

Gips, C., Hofstedt, P., and Wysotzki, F. (2002). Spatial Inference – Learning vs. Constraint Solving. In Jarke, M., Köhler, J., and Lakemeyer, G., editors, *Proceedings of the 25. German Conference on Artificial Intelligence, Sept. 16-20, 2002, Aachen, Germany*, LNAI. Springer. to appear.

Guesgen, H. W. (1989). Spatial Reasoning Based on Allen's Temporal Logic. Technical Report TR-89-049, ICSI, Berkeley, Cal.

Hernández, D. (1994). *Qualitative Representation of Spatial Knowledge*, volume 804 of *LNAI*. Springer.

Hofstedt, P. (2000). Cooperating Constraint Solvers. In *Sixth International Conference on Principles and Practice of Constraint Programming - CP*, volume 1894 of *LNCS*. Springer.

Johnson-Laird, P. N. (1983). *Mental Models: Towards a Cognitive Science of Language, Inference and Consciousness*. Cambridge University Press, Cambridge.

Jörding, T. and Wachsmuth, I. (1996). An Antropomorphic Agent for the Use of Spatial Language. In *Proceedings of ECAI'96-Workshop on Representation and Processing of Spatial Expressions*, pages 41–53.

Müller, W. and Wysotzki, F. (1997). The Decision-Tree Algorithm CAL5 Based on a Statistical Approach to Its Splitting Algorithm. In Nakhaeizadeh, G. and Taylor, C. C., editors, *Machine Learning and Statistics - The Interface*, pages 45–65. Wiley.

Röfer, T. (1997). Routemark-based Navigation of a Wheelchair. In *Third ECPD International Conference on Advanced Robotics, Intelligent Automation and Active Systems*, Bremen.

Schulmeister, B. and Wysotzki, F. (1994). The Piecewise Linear Classifier DIPOL92. In Bergadano, F. and de Raedt, L., editors, *Proceedings of the European Conference on Machine Learning*, volume 784 of *LNAI*, pages 411–414, Berlin. Springer.

Wiebrock, S. (2000). Anwendung des Lernalgorithmus CAL5 zur Generierung von Depiktionen und zur Inferenz von räumlichen Relationen. Technical Report 2000-14, ISSN 1436-9915, TU Berlin.

Wiebrock, S., Wittenburg, L., Schmid, U., and Wysotzki, F. (2000). Inference and Visualization of Spatial Relations. In Freksa, C., Brauer, W., Habel, C., and Wender, K., editors, *Spatial Cognition II*, volume 1849 of *LNAI*. Springer.

Wiebrock, S. and Wysotzki, F. (1999). Lernen von räumlichen Relationen mit CAL5 und DIPOL. Technical Report 99-17, TU Berlin.